The New York Times

LOOKING FORWARD

Extraterrestrials and U.F.O.s

THE NEW YORK TIMES EDITORIAL STAFF

Published in 2020 by New York Times Educational Publishing in association with The Rosen Publishing Group, Inc.
29 East 21st Street, New York, NY 10010

Contains material from The New York Times and is reprinted by permission. Copyright © 2020 The New York Times. All rights reserved.

Rosen Publishing materials copyright © 2020 The Rosen Publishing Group, Inc. All rights reserved. Distributed exclusively by Rosen Publishing.

First Edition

The New York Times
Alex Ward: Editorial Director, Book Development
Phyllis Collazo: Photo Rights/Permissions Editor
Heidi Giovine: Administrative Manager

Rosen Publishing
Megan Kellerman: Managing Editor
Julia Bosson: Editor
Greg Tucker: Creative Director
Brian Garvey: Art Director

Cataloging-in-Publication Data
Names: New York Times Company.
Title: Extraterrestrials and U.F.O.s / edited by the New York Times editorial staff.
Description: New York : New York Times Educational Publishing, 2020. | Series: Looking forward | Includes glossary and index.
Identifiers: ISBN 9781642822663 (library bound) | ISBN 9781642822656 (pbk.) | ISBN 9781642822670 (ebook)
Subjects: LCSH: Life on other planets—Juvenile literature. | Extraterrestrial beings—Juvenile literature. | Unidentified flying objects—Juvenile literature.
Classification: LCC QB54.E987 2020 | DDC 576.8'39—dc23

Manufactured in the United States of America

On the cover: The dome-shaped store at Judy Messoline's U.F.O. Watchtower in Hooper, Colo. Many of the visitors to Messoline's Watchtower have experienced a shudder in the fabric of the ordinary that has drawn them to an otherwise empty spot in south central Colorado; Kevin Moloney for The New York Times.

Contents

7   Introduction

CHAPTER 1

Aliens and the Cultural Imagination

10   Why We Keep Dreaming of Little Green Men **BY GEORGE JOHNSON**

14   Taking U.F.O.'s for Credit, and for Real **BY MICHAEL DECOURCY HINDS**

17   U.F.O. Believers and Debunkers Thrive on the Web
        BY PATRICK J. LYONS

20   Lonesome Highway to Another World? **BY STEPHEN REGENOLD**

26   Pit Stop for U.F.O.'s, and Humans Who Love Them **BY KIRK JOHNSON**

30   The Truth Is Out There **BY JAMES RYERSON**

34   Flying Saucers and Other Fairy Tales **BY ROSS DOUTHAT**

37   U.F.O.s: Is This All There Is? **BY DENNIS OVERBYE**

42   They've 'Seen Things' **BY ROZETTE RAGO**

CHAPTER 2

Aliens and the U.S. Government

52   Visitors From Outer Space, Real or Not, Are Focus of Discussion in Washington **BY ANDREW SIDDONS**

56 C.I.A. Acknowledges Area 51 Exists, but What About Those Little Green Men? BY ADAM NAGOURNEY

60 Hillary Clinton Gives U.F.O. Buffs Hope She Will Open the X-Files BY AMY CHOZICK

64 2 Navy Airmen and an Object That 'Accelerated Like Nothing I've Ever Seen' BY HELENE COOPER, LESLIE KEAN AND RALPH BLUMENTHAL

67 Glowing Auras and 'Black Money': The Pentagon's Mysterious U.F.O. Program BY HELENE COOPER, RALPH BLUMENTHAL AND LESLIE KEAN

75 On the Trail of a Secret Pentagon U.F.O. Program BY RALPH BLUMENTHAL

CHAPTER 3

The Search for Extraterrestrial Life

78 The Intelligent-Life Lottery BY GEORGE JOHNSON

83 Should We Keep a Low Profile in Space? BY SETH SHOSTAK

87 The Flip Side of Optimism About Life on Other Planets BY DENNIS OVERBYE

92 Yes, There Have Been Aliens BY ADAM FRANK

96 Twinkle, Twinkle Little Trappist BY THE NEW YORK TIMES

98 A New Exoplanet May Be Most Promising Yet in Search for Life BY DENNIS OVERBYE

101 'Aliens' Asks: If the Universe Is So Vast, Where Is Everybody? BY JENNIFER SENIOR

105 Greetings, E.T. (Please Don't Murder Us.) BY STEVEN JOHNSON

128 An Interstellar Visitor Both Familiar and Alien BY DENNIS OVERBYE

132 A Nearby Earth-Size Planet May Have Conditions for Life
BY KENNETH CHANG

135 A Large Body of Water on Mars Is Detected, Raising the Potential for Alien Life BY KENNETH CHANG AND DENNIS OVERBYE

CHAPTER 4

Believers

140 Betty Hill, 85, Figure in Alien Abduction Case, Dies BY MARGALIT FOX

143 Philip Klass, 85, Debunker of Claims of Flying Saucers, Dies
BY DOUGLAS MARTIN

145 When an Astronaut Believes in Aliens BY MIKE NIZZA

147 Budd Hopkins, Abstract Expressionist and U.F.O. Author, Dies at 80 BY MARGALIT FOX

151 John Billingham, Seeker of Extraterrestrials, Dies at 83
BY WILLIAM YARDLEY

154 Ionel Talpazan, Whose U.F.O. Art Had Sightings All Over, Dies at 60 BY WILLIAM GRIMES

157 The Woman Who Might Find Us Another Earth BY CHRIS JONES

171 Winston Churchill Wrote of Alien Life in a Lost Essay
BY KIMIKO DE FREYTAS-TAMURA

175 Ursula Marvin, Geologist of the Extraterrestrial, Dies at 96
BY RICHARD SANDOMIR

CHAPTER 5

Encounters

179 Despite Lack of Data From Pilots and Officials, Reports of UFO Sightings Are Many and Widespread BY WALTER SULLIVAN

186 A U.F.O., or Is There an Explanation? BY ALBERT J. PARISI

189 U.F.O. Hits Congestion at O'Hare, Turns Back BY TOM ZELLER JR.

191 Formations in China Desert Are Still a Mystery BY J. DAVID GOODMAN

194 Bright Lights, Strange Shapes and Talk of U.F.O.'s
BY JONAH ENGEL BROMWICH

197 People Are Seeing U.F.O.s Everywhere, and This Book Proves It
BY RALPH BLUMENTHAL

201 Was a Tiny Mummy in the Atacama an Alien? No, but the Real Story Is Almost as Strange BY CARL ZIMMER

205 A Radar Blip, a Flash of Light: How U.F.O.s 'Exploded' Into Public View BY LAURA M. HOLSON

211 Glossary
213 Media Literacy Terms
215 Media Literacy Questions
217 Citations
222 Index

Introduction

ON DEC. 27, 2018, a mysterious electric-blue light emerged above the skyline of New York City. Within minutes, the lights had been documented by hundreds of concerned New Yorkers, who shared photographs of the eerie light and wondered where it could have come from. Many noted that the color seemed unnatural. The Internet lit on fire with hypotheses: Could this be a sign of alien life?

Ultimately, officials revealed that the blue lights were caused by a power plant in Queens, but the furor the lights caused is indicative of a much larger fascination with extraterrestrial life. The idea of life outside Earth animates countless tabloid articles as well as thousands of movies and nearly the entire genre of science fiction. Areas in the United States rumored to have U.F.O. sightings attract hordes of visitors each year, hoping for an encounter of an alien kind. Communities of believers gather across the country, scouring news stories for hints of conspiracy and pressing the United States to reveal details of its most classified programs. Whether or not proof exists, the impact of aliens is felt among us all.

Belief in alien life is the rare interest that unifies conspiracy theorists and astrophysicists. For those who study stars and planets, the search for alien life represents the next frontier of space research. The more we begin to understand the universe, the more pressing the question becomes: Are we alone?

The articles in this book speak to the scope and significance of that question. Journalists have traveled to remote parts of the American West, where whole communities have formed to investigate the possibility of alien life on Earth and interested tourists have followed. Reporters have embedded themselves in communities of believers and

JED MCGOWAN

interviewed men and women who claim to have encountered aliens. And science writers explore rumors of possible U.F.O. sightings and look into the science that explains them away.

At the same time, some of these articles explore the science that suggests that we may in fact be close to identifying life outside of Earth. Scientists have been searching our solar system as well as other parts of the galaxy. In recent years, researchers have discovered evidence of water on Mars, suggesting that at some point there may have been actual life there. In 2017, a discovery of numerous Earth-size planets that showed the potential to host life spurred a flurry of interest and questions of what humans might do if we found life forms there. The SETI (Search for Extraterrestrial Intelligence) Institute, founded by Carl Sagan and Frank Drake, has received millions of dollars in funding to support their initiatives.

In fact, it's not just scientists who think that meeting aliens might be possible. A 2017 report by The New York Times showed the U.S. government's interest in possible U.F.O. sightings. As conspiracy theorists have long claimed, secret governmental and military agencies have explored technology that might make an encounter with extraterrestrial life possible.

Finally, this book also considers the future of our search for aliens: If there is life outside of planet Earth, then how would we communicate with it or them? Do we trust ourselves as humans to approach life outside Earth with generosity? And as climate change and industrialization begin to raise doubts about the long-term future of our planet, the search for life-providing planets has taken on a new urgency. Not only do we seek evidence of other civilizations, but we also look for the possibility of extending our own.

In many ways, the hunt for extraterrestrials and U.F.O.s is just human instinct. What is more frightening than the possibility of encountering a hostile alien civilization is the idea that, in the wide universe, we are wholly alone.

CHAPTER 1

Aliens and the Cultural Imagination

Humans have long had an obsession with the possibility of life on other planets. Reports of extraterrestrial sightings have been standard tabloid fodder, and U.F.O.s are a staple of science fiction. While the image of little green men emerging from a flying saucer has become an outmoded one, aliens continue to fascinate and hold an influence on popular culture. The articles in this chapter examine the obsession humans have with life outside of Earth and the hold that aliens have on the cultural imagination.

Why We Keep Dreaming of Little Green Men

OPINION | BY GEORGE JOHNSON | MAY 13, 2016

CONSPIRACY THEORISTS TEND to cluster at the right and left of the political spectrum, so perhaps Hillary Clinton will attract at least a few voters from both the Trump and Sanders camps with her recent pledge to release documents about Area 51, the top secret military base in Nevada.

Some of her critics have been dubious, suspecting that her vow, made on forums like Jimmy Kimmel's late-night talk show, was intended to distract attention from her emails and Goldman Sachs speeches. But among the electorate, the antennas of some U.F.O. seekers must have perked up — polls have found that more than a third of Americans are believers — when they heard a presidential candidate actually talking about Area 51.

Suspecting that deep secrets are hidden there — in the form of captured aliens (dead or alive), crashed extraterrestrial spaceships and futuristic weaponry — U.F.O. die-hards have long pushed the government to come clean about the facility, which was established in 1955 as an annex of the Atomic Energy Commission's Nevada Test Site. The official story — that Area 51 has been used for secret activities like testing the prototype of the U-2 spy plane and other experimental aircraft — seems to them like a cover-up, a suspicion fed by the government's refusal to acknowledge the place's existence until the release of a classified report in 2013.

There was no mention of extraterrestrials, of course. What else are our leaders, who may be aliens themselves, going to say?

The number of people fixated on Area 51 to the exclusion of other issues is probably not enough to swing more than a precinct here and there. But the fascination with alien cover-ups taps into a deeper vein. Maybe it's different for the 0.1 percent at the top of the pyramid (you with your shining eye on the back of the dollar bill), but for most of us the world is a confusing, complicated, mind-numbing place over which we feel a dismaying lack of control.

Sometimes one suspects that a piece of the puzzle must be missing, or dangling cruelly beyond our reach. You can either muddle along without it, as most of us try to do, or put your mind into hyperdrive, making connection after connection and piecing together a hidden order — a conspiracy so immense that it threatens to be more convoluted and complex than what it seeks to explain.

The truth is out there or in there. Open the gates to the inner sanctum — whether it's the Vatican, the Federal Reserve, the Masonic lodges or Area 51 — and suddenly everything will make sense.

Masters of the craft (there is no good word, at least in English, for conspiracy theorizing) could be found in the 18th century, busily writing tracts and tomes concluding that the French Revolution was plotted by Freemasons working with the Bavarian Illuminati. The hypothesis quickly grew to include the Rosicrucians, the Knights

Templar, the Cathars and ancient Egyptian religious cults — ingredients that Dan Brown made lucrative use of in "The Da Vinci Code." All were said to be players in a secret world history that had been unfolding backstage for centuries, while the masses were distracted like children with shadow plays.

From Europe this spider web of memes spread to the United States, where the existence of Masonic lodges led to suspicions of an Illuminati plot to surrender the country to France, a land teeming with Enlightenment philosophers with godless beliefs and cosmopolitan ways — the original secular humanists.

Protected in its cocoon, this style of thinking — the "paranoid style," the historian Richard Hofstadter memorably called it — was carried intact into modern times, with the nexus of evil moving to the Soviet Union, the Trilateral Commission and other suspected agents of One World Government. The next step was surely surrender to the Galactic Empire. No wall — it would have to be a planetary shell — could stop the ultimate aliens. "The Day the Earth Stood Still" was both a warning and a diversion. No wonder there are strange lights in the sky.

So go the enduring themes in the literature of paranoia. From early on, the world's secret rulers were said to possess extraordinary powers, supernatural and quasi-scientific. In the 18th century these included mesmerism, which later gave way to telepathy and then mind control through cellphone waves and chemtrails, spewed by aircraft launched, perhaps, from Area 51.

President Obama's real birth certificate may be there along with a new cache of climategate emails and the refrigerated room where John F. Kennedy is kept alive, hooked to electrodes, robotically enhanced and ready to take charge someday as the bionic president.

It's easy to get carried away, your reptilian brain fueling your cybernetic cerebrum, as click, click, click, you feel the pieces snapping together. The followers of Lyndon LaRouche, the world-class conspiracy theorist who runs periodically for president, propound a cracked and erudite worldview that has included conspirators like Aristotle, John Maynard Keynes, Werner Heisenberg and Timothy Leary — all linked through an internal logic that makes, for its believers, a scary kind of sense.

It takes great mental powers to construct these intricacies no matter how crazy they are. Conspiracy theorists are not stupid people. Given a different turn in life, some might have made good superstring theorists.

"The higher paranoid scholarship is nothing if not coherent," Hofstadter wrote in "The Paranoid Style in American Politics." "In fact the paranoid mind is far more coherent than the real world."

Other scholars have found that adherents of one conspiracy theory are likely to believe in others. They are good — maybe too good — at making connections. Maybe the phenomenon is neurological, with synapses packed so densely that the brain is driven to see way more order than can possibly be there.

There are, of course, sensible reasons to believe in extraterrestrial life. Just last week, evidence for more than 1,200 new planets was reported. A fraction of those may harbor life, and a fraction of a fraction may have produced intelligent creatures. Exploring the universe, some may have caught sight of Earth and even landed. But probably not.

If Mrs. Clinton is serious and releases files on Area 51 — those, she adds, that do not breach the country's security — the result will probably be anticlimactic. More fuel for the conspiracy theorists.

GEORGE JOHNSON, who writes the Raw Data column for Science Times, is the author of "Architects of Fear: Conspiracy Theories and Paranoia in American Politics."

Taking U.F.O.'s for Credit, and for Real

BY MICHAEL DECOURCY HINDS | OCT. 28, 1992

PHILADELPHIA, OCT. 27 — Every fall semester at Temple University, Prof. David M. Jacobs exposes himself to ridicule by teaching a course entitled "Unidentified Flying Objects in American Society." This year he made himself more vulnerable by asserting in a book that extraterrestrials are among us and doing vile things.

"We have been invaded," he writes. "They want human sperm and eggs."

Such theories, he admits, make most people laugh or call him an embarrassment to the university. But many faculty members strongly support him, if only because of their commitment to academic freedom, and students sign up for his class in droves and give it superlative ratings in questionnaires.

Professor Jacobs, a historian, graduated from U.C.L.A. and earned his Ph.D. in history from the University of Wisconsin, where he wrote his doctoral thesis about the U.F.O. phenomenon. Having spent nearly a third of his 50 years on this planet separating U.F.O. fact from U.F.O. fiction, he is the first to admit his findings sound incredible. A corkboard in his cluttered Temple office, for example, features clippings from supermarket tabloids with headlines like "Alien Kidnappers Seek Human Blood and Body Parts, Warn Experts."

So what is the difference between aliens seeking blood and aliens seeking sperm and eggs? Research, says Professor Jacobs. And his book, "Secret Life: Firsthand Accounts of UFO Abductions" (Simon & Schuster, 1992), has a glowing introduction by Dr. John E. Mack, a professor of psychiatry at Harvard Medical School. Professor Jacobs had "made his case well," Dr. Mack writes.

Using hypnosis, Professor Jacobs interviewed about 60 people who say they were kidnapped and sexually exploited by aliens in several hundred encounters. Their stories are amazingly similar. They tell of "small beings" and "taller beings" from 2 feet to 5 feet tall with big heads, huge eyes and leathery gray skin. They do not talk but communicate telepathically, the people say.

The "beings," they say, kidnap men and women at night or at other times when the victims will not be missed, then put them into a trance, give them medical examinations, extract sperm or eggs, and sometimes implant or remove alien-human embryos.

Millions of Americans may have been abducted, Professor Jacobs theorizes, but most of them know only a residual feeling of anxiety and depression akin to post-traumatic stress disorder. "I want to be wrong," he said of his theory, adding he would "weep with joy" if there were no aliens around.

Professor Jacobs, who teaches a variety of history courses, has taught the one-semester U.F.O. course since 1977. The course, which attracts about 75 students, concludes with the study of his research on abductions.

"I am an easy target for ridicule," Professor Jacobs said. "But this is a serious course about a serious subject, and students take it seriously and get serious grades."

During a recent class, students paid rapt attention as Professor Jacobs described how a scholarly committee in the late 1960's was commissioned by the Air Force to investigate reported U.F.O. sightings. Instead, he said, committee members conspired to discount the validity of the sightings.

"I love it," Gregory Webb, a sophomore, said at the end of class.

Joe Clemens, a senior, said, "I absolutely believe there are U.F.O.'s."

But Lou Teller, a senior, was skeptical. "I feel he is trying to sell us his view on U.F.O.'s because he discounts everyone else's work," he said.

Among themselves, some professors have questioned whether Temple's curriculum should include a three-credit course discussing supposed sexual abuse by aliens. But academic freedom has triumphed.

"I don't see David proselytizing, and as long as that is the case, I will continue to support him," said James W. Hilty, chairman of the history department. He recently nominated Professor Jacobs for promotion, to full professor of history from associate professor on the basis of "Secret Life" and numerous articles on U.F.O.'s. "It's an important body of work, and he has shown courage in pursuing it," Professor Hilty said.

Other faculty members think Professor Jacobs's book and course are nonsense. "Faculty members talk about it a lot and say it's really oddball stuff," said Prof. Philip R. Yannella, director of the department of American studies. "But I wouldn't want to be in the position of canceling the course just because it advocated a strange position."

Prof. Julia A. Ericksen, until recently the university's acting provost with responsibility for curriculum, said: "The most troubling complaints were two letters saying that the people Professor Jacobs had interviewed for his book had to be mentally ill, and that it was irresponsible for him to encourage their fantasies rather than getting them treatment."

But then again, Professor Ericksen said, "People didn't believe Copernicus either."

U.F.O. Believers and Debunkers Thrive on the Web

BY PATRICK J. LYONS | JUNE 30, 1997

MENTION U.F.O.'S AND the Internet in the same sentence and eyes roll.

With some justice, too, since the Internet has proved to be an ideal outlet for crackpotism of every flavor. Conspiracy theorists and true believers have flooded the World Wide Web with rant, screed, pseudoscience, paranoid fantasy, arcane enigmata, fifth-hand rumor and just plain fiction on the subject of unidentified flying objects. The work of more serious people who study reports of supposed alien visitation and the widespread popular belief in unidentified flying objects can easily be lost in all the panting and malarkey.

Ground zero for ufologists, serious and wild-eyed alike, has long been a stretch of scrubland outside Corona, N.M., where a rancher found a bit of puzzling debris in July 1947. Initial reports in the local newspaper in Roswell, the nearest sizable town and the home of an Air Force base, called it a crashed flying saucer, and five decades later people still believe that is what it was.

Roswell, Hangar 18 and Area 51 (top-secret spots in Nevada where bits of the ship and its dead occupants supposedly were taken) have become part of American folklore, notwithstanding 50 years of Air Force insistence that all anyone ever found were parts of a secret high-altitude research balloon.

The vast edifice of flying saucer and alien abduction lore built up over the years is actively trafficked on the Internet, at Web sites like those of Uforia, the National UFO Reporting Center, the Mutual UFO Network or the International UFO Museum and Research Center in Roswell.

Fittingly, then, the latest and most comprehensive debunking of the supposed Roswell incident to come from the Air Force is also available on the Web. "The Roswell Report: Case Closed," published last week, is posted on the Air Force Library site, photos and all, along with its previous attempt at a stake through the monster's heart, the 1994 "Roswell Report."

Naturally, people who believe that the Government covered up a crashed alien ship for 50 years will be inclined to dismiss these reports as just more covering-up. But the Air Force Library is far from the only place to find cold water thrown on overheated imaginings of extraterrestrial desert sightseeing tours gone very, very wrong.

The leaders in general-purpose skepticism have long been James Randi and the Committee for Scientific Investigation of Claims of the Paranormal, or Csicop, which he helped to found. Mr. Randi was once the Amazing Randi, a performing illusionist, but his impatience with the psychic pretensions of fellow performers led him to devote more and more time to exposing those he considers hucksters and charlatans, including purveyors of dubious space-alien stories.

Mr. Randi's Web site mirrors his witheringly sarcastic take on spoon-benders, spirit mediums, mind readers and their too-trusting marks, and includes contributions from the like-minded team of Penn and Teller. The site also gives the ground rules for a prize that Mr. Randi has long offered for compelling proof of anything paranormal — as yet unclaimed, and grown now to more than $1.1 million.

(Mr. Randi's attitude has been so derisive for so long that he has become a lightning rod for the fury and vitriol of the passionately credulous. A typically incoherent attack on Mr. Randi's prize offer can be found on the Web page of Bruce D. Kettler, apparently in defense of a clairvoyant named Ed Dames.)

Csicop's site gathers together links to carefully reasoned analysis of all sorts of implausibilities, including the Roswell incident and many other aspects of the U.F.O. phenomenon. The site features highlights from Csicop's magazine, The Skeptical Inquirer, which dissected matters Roswellian in its July-August 1995 issue. An analogous antipodean group, Australian Skeptics, maintains another reality-check site, as do scientists' groups around the world. And the writings of professional U.F.O. skeptics like Philip Klass are widely linked on the Web.

Individual bits of Roswell "evidence" come in for their own custom debunkings on the Web. An article on The Albuquerque Journal's site finds a simple terrestrial explanation for a "mystery metal" fragment that made its way into the Roswell Museum. The much-ballyhooed "alien autopsy film" and its checkered history get the treatment on the Manikin Who Fell to Earth page by James Easton; assertions that that Eastman Kodak "authenticated" the film are disposed of neatly on Paul Fuller's page. A competing set of still photos of an "alien autopsy" bought by the publisher of Penthouse Magazine, Bob Guccione, are called not just a hoax but a retreaded one on Parascope's site.

There are even how-to sites for U.F.O. hoaxers. Truly Dangerous offers advice on fabricating an alien "corpse" for autopsy purposes, along with links to lots of Roswell-debunking sites. Whipping up a mysterious floating light in the sky is easy once you have seen Roel van der Meulen's directions in the Project Galactic Guide. Creators of Britain's "mysterious" crop circles fess up on the Circlemakers site.

A longtime ufologist, James W. Moseley, offers a peek behind the curtain on his Saucer Smear site, with gossip about his peers that he calls "shockingly close to the truth."

Unfortunately, a well-known, unintentionally hilarious 52-question test to see if you have ever been abducted by aliens appears to have vanished from the Web recently, leaving behind a lot of dead-end links to it. But you can still determine whether you yourself are really an alien by taking the Sleeping ET Quiz. One sure sign: "People say you're naive."

Lonesome Highway to Another World?

BY STEPHEN REGENOLD | APRIL 13, 2007

A MOMENT BEFORE the sonic boom hit his trailer, Joerg Arnu's UHF radio scanner crackled to life. "Cylon 1, got you on radar," said a voice just barely perceptible through the static.

And then — badamm-booom! — the whole trailer shook in a shockwave, and Mr. Arnu jumped, a big plexiglass window reverberating as a jet streaked overhead and through the sky.

"That was probably an F-16," Mr. Arnu said, peering out the window and squinting into the sun. A telephoto lens sat on a countertop nearby.

"They're testing a new weapon lately, and a laser system to shoot down missiles," he said.

From his trailer in the town of Rachel, Nev., Mr. Arnu is less than 10 miles from an unmarked military boundary, beyond which the top-secret Air Force base known as Area 51 sits on a dry salt flat guarded by big arid mountains and bleak desert on all sides.

To the east, tracking past Rachel in two asphalt lanes, Nevada State Route 375 bisects a wide basin, coursing northbound before disappearing into a haze of nothingness beyond.

This is Alien Country, where more U.F.O.'s are sighted each year than at any other place on the planet, at least according to Larry Friedman of the Nevada Commission on Tourism.

A sign outside Rachel declares Nevada State Route 375 to be the Extraterrestrial Highway, the name given to the road in 1996. Renaming the road, the tourism commission had hoped at the time, would draw travelers to the austere and remote reaches of south-central Nevada, where old atomic bomb test sites, secret Defense Department airstrips and huge, sequestered tracts of military land create a marketable mystique.

Oh, and don't forget the flying saucers.

ISAAC BREKKEN FOR THE NEW YORK TIMES

The Extraterrestrial Highway, one of the most desolate roads in the country, has more reported U.F.O. sightings than any other road in the country.

"People now come all the way from Japan to see what this place is about," Mr. Friedman said.

Indeed, on a recent Wednesday afternoon, after a two-hour drive up from Las Vegas through the utter emptiness of Lincoln County, the first tourist I met on the Extraterrestrial Highway was from Yamaguchi Prefecture in southwest Japan.

"We came for an alien souvenir," said Shihgo Miyamoto, 29, who was holding his wife, Yoko, both shivering in the high-desert wind. I took their picture under an official Nevada Department of Transportation Extraterrestrial Highway sign, a sprawl of trailer homes in the background.

"So cold, so empty," said Mr. Miyamoto, looking to the desert beyond his rental car.

South-central Nevada is, by and large, a vast wasteland, scrubby and unpopulated, dotted with dry lakes, streaked with tan rocky

ISAAC BREKKEN FOR THE NEW YORK TIMES

A helpful sign at the Little A'Le'Inn.

peaks, ravines and wide alluvial plains. Government land is ubiquitous. Cattle guards rumble under tires on the barren highways, which cut through sand and open range. To drive the Extraterrestrial Highway — a route that snakes northwest for 98 empty miles, intersecting no other major roads — is to drive one of the most desolate stretches of pavement in the country. Gasoline is unavailable for its entire length. R.V.'s cannot hook up in Rachel, the only town on the road.

According to the Nevada Department of Transportation, an average of about 200 cars drive some portion of the Extraterrestrial Highway every day, making it one of the state's least traveled routes.

On my midday drive up the highway in February, I saw only six other vehicles.

Coming north from the town of Alamo, where I stayed overnight in a cabin, the Extraterrestrial Highway began as an innocuous flat road through scrubby highlands. A mile or so in, a large silver Quonset hut announced itself as the New Alien Research Center, but its driveway was gated, so I drove on by.

The road bobbed through a Martian landscape, red valleys raked with lines, flat expanses of gravel and dead shrubs, all ringed by hulking mountains of stratified stone.

A hawk hung high in the air. Joshua trees reached for the sun, their bristled bunches aglow, seemingly illuminated from within.

But soon I forgot about the nature and started looking for U.F.O.'s. A sign warned of low-flying aircraft. Contrails streaked the blue yonder ahead.

In Rachel, 40 minutes into the drive, I stopped at the Little A'Le'Inn (pronounced Little Alien), a bar and restaurant, which sells extraterrestrial-themed mixed drinks alongside self-published books like "The Area 51 & S-4 Handbook." Its walls were covered with sun-faded photographs featuring aliens, glowing orbs and obelisks zooming through clouds.

The bartender was polishing a glass, standing near a man slumped over a drink, when I approached to inquire about area attractions. "You should talk to Pam," the bartender said, pointing to a woman standing near the door.

And so I was introduced to Pam Kinsey, the first of several residents I met eager to talk about Rachel, and Area 51, and government sensors hidden in sand, and glowing dots hovering on high.

But Ms. Kinsey, 42, who has lived in the area for almost two decades, is not herself an ardent alien believer.

"We have a military base next door that can explain a lot of the lasers and other weird things," she said.

Ms. Kinsey said that only a couple of Rachel's 75 or so residents talk about seeing saucers and little green men. The tourists — whom she confirmed come from all over the world — are often the only extraterrestrial seekers found in Rachel.

"There are conventions held in town, and the alien people like to come here and congregate," she said.

(On Memorial Day weekend, May 25 to 27, the Little A'Le'Inn will play host to its sixth annual U.F.O. Friendship Campout, which includes seminars, book signings and nightly sky watches led by a "certified U.F.O. Investigator.")

DeWayne Davis, a 72-year-old retired Air Force engineer who came to the Little A'Le'Inn for dinner, said he has seen saucers in the area, including a glowing craft that hovered at high altitude before tracing a rectangular pattern in the night sky.

"It was at 55,000 feet or higher," he said. "And it emitted an orange sodium-vapor color, not the xenon glow you'd usually see."

Mr. Davis, who said he worked at a military installation in Roswell, N.M., during the mid-'50s, moved to central Nevada in 1997 for the clean air, the solitude and the scenery. He now lives in a trailer a couple of blocks off the Extraterrestrial Highway. The frequent sonic booms of test planes breaking the sound barrier overhead are music to his ears, he said.

Outside the Little A'Le'Inn, I walked a few dusty blocks to take in the sights around town, including an ad hoc air-traffic control tower draped in camouflage netting.

Jets streaked overhead, silent at high altitude, blazing west toward a setting sun.

Before leaving the area, I drove out of town a mile to find Mr. Arnu, a 45-year-old software developer from Las Vegas who keeps a trailer parked on some land he purchased in 2003 as a retreat from the city. Mr. Arnu, a native of Germany who runs www.dreamlandresort.com, a popular Web site on Area 51, said that he files a Freedom of Information Act petition each year to procure dates and times of major military testing periods. "That's when all the action happens," he said.

My visit to the area coincided with Red Flag, the name Mr. Arnu gave a period in mid-February when military exercises out of nearby Nellis Air Force Base send a proliferation of jets into the air.

"Earlier today, I saw British Tornados, American F-22s and an Australian F-111," said Mr. Arnu, who hikes the hills around town to photograph supersonic planes. He lives for the simulated dogfights that take place in the air above the Extraterrestrial Highway.

Like most local people I met, Mr. Arnu thinks the Nevada Commission on Tourism's fixation with aliens is a bit silly.

"I'm a plane-spotter," he said. "I have no real belief in the alien stuff."

Driving alone later that night, the Extraterrestrial Highway a dark winding lane in my headlights, I wasn't sure what to think. On a mountain pass 20 minutes from town, I parked my car and shut off the engine, an inky abyss closing in from all sides.

Stars packed the deep velour above, hundreds of thousands of humming and twinkling little jewels. A blinking red dot dipped behind a mountain in the distance.

I waited, searching the sky.

But nothing moved, nothing came, and I started to get cold. My red dot was just a jet, probably descending to a landing in Las Vegas 100 miles to the south.

The desert wind howled in a valley below. It was black and cold. On the Extraterrestrial Highway, I was all alone.

Pit Stop for U.F.O.'s, and Humans Who Love Them

BY KIRK JOHNSON | NOV. 25, 2010

HOOPER, COLO. — "I like humans, they're fun," Judy Messoline said as she showed a visitor through her vortex garden, which psychics have said contains not just one, but two separate portals to a parallel universe.

Many of the humans who come to Ms. Messoline's U.F.O. Watchtower, hard by the dueling vortexes, may be fun, but they are also wounded. About 95 percent, by her estimate — and she makes a point of asking — have experienced something, a shudder in the fabric of the ordinary, the sighting of an unidentified flying object that to one degree or another has haunted them and drawn them to this otherwise empty spot in south-central Colorado. Having fun in thinking about extraterrestrials, she said, is usually bound up with something deeper right here on the home planet.

"The world needs a place where people can go to talk about their experiences and not be laughed at," she said.

People do laugh here. One of Ms. Messoline's principles in building the Watchtower a decade ago, in an attempt to raise cash as her cattle ranch collapsed in economic ruin, was that U.F.O.-spotting should be a hoot, and whenever possible, a party.

"The best sightings have been when people are just out enjoying the evening," she said. Fifty-nine events — lights that move erratically or, during the day, objects that defy explanation in shape or movement — have been witnessed from the tower since 2000, Ms. Messoline said, sometimes by dozens of people at the same time.

No one knows the count before that, since no local institution existed for counting. Many residents, though, say the San Luis Valley, just north of the New Mexico state line, has been a hotspot for decades. U.F.O. reports reach all the way back to the early settlements of the 1600s, with a particularly noted wave in the late 1960s.

KEVIN MOLONEY FOR THE NEW YORK TIMES

Judy Messoline and her U.F.O. Watchtower in Colorado's San Luis Valley. "The world needs a place where people can go to talk about their experiences and not be laughed at," she said.

The turmoil of modern life is also in evidence near the tower, at the house once occupied by Ms. Messoline's son and his family, now vacant and in foreclosure since the couple's divorce.

"Broke my heart," she said. Adding to the pain, she said, is that the house will probably never sell. "Who wants to live next to a U.F.O. Watchtower?" she said.

Truth be told, the Watchtower — really just a framed metal platform perhaps 10 feet off the ground — is not much of a moneymaker at $2 a head for admission. Ms. Messoline, 65, a former housecleaner from the Denver area who moved to Hooper in the mid-1990s, still needs the paycheck from Miss Deb's, a convenience store down the road, identified by the giant chicken out in front, to make ends meet.

But that is the interconnection of a lot of things in Hooper, a dot of perhaps 100 souls in a vast and lonely place. Harsh realities in economics and climate — high poverty rates and brutal winters — are

KEVIN MOLONEY FOR THE NEW YORK TIMES

Visitors leave behind gifts for any surprise guests.

interlaced with vistas of breathtaking beauty and a local culture that has long prized and cultivated the offbeat.

Ms. Messoline furthered that spirit by encouraging visitors to leave something in her vortex garden. One recent offering: a two-foot-tall Superman doll with one hand extended, holding a bottle of hot sauce, perhaps in greeting or in supplication.

Another visitor left a primer for extraterrestrials who might find themselves confused about human tableware. A folding knife-and-spoon was marked with text and helpful arrows pointing in the direction of each object: "This is a knife and a spoon, alien," it said.

Even the winds are strange. One corner of the San Luis Valley, banked on all sides by mountains, somehow became a collecting spot for blown sand over the past few thousand years, since the drying up of an ancient lake bed. The result: a little bit of the Sahara in Colorado at Great Sand Dunes National Park and Preserve, about 15 miles from here.

The sky, with barely a town to break the landscape, is black at night — a riot of stars not visible from the big city — and huge at all hours. And people here are used to being out and aware of their surroundings, which makes them perhaps more likely than city folk to see things in the great Out There.

"There's not a lot of activity, so people have more opportunity to be watching what's around them," said JoDene Newmyer, 64, who works with Ms. Messoline at the convenience store.

Ms. Newmyer's own U.F.O. story — and most people here seem to have one — occurred on the Friday morning of Memorial Day weekend, 1972. She was driving her daughter to the baby sitter at 7 a.m. when she stopped cold at the sight of a huge angular silver object just above the horizon.

"Flying saucer? I will not say that," Ms. Newmyer said. "But unidentifiable it definitely was, because I've never seen anything like it."

Ms. Messoline says the years of scanning the sky and of meeting people who are drawn to her and her tower have changed her.

She decided recently to put the patch of ground under the tower and the vortex garden in her will, donating it to a U.F.O. research group in Denver to continue the work, or the fun, after she's gone, even though she knows that a tower in perpetuity will probably doom any chance of a sale of her son's former home.

The Truth Is Out There

REVIEW | BY JAMES RYERSON | SEPT. 23, 2016

IN OCTOBER OF 1995, two Swiss astronomers announced a major discovery: They had detected, about 50 light-years from Earth, a planet orbiting a sun-like star. Scientists had long imagined there were such planets outside our solar system, but never before had one been confirmed. This extrasolar planet, or "exoplanet," appeared to be a Jupiter-like ball of gas and liquid with a blistering atmospheric temperature of 1,700 degrees Fahrenheit. In the two decades since, astronomers have detected more than 3,000 exoplanets. No one has yet found a replica of Earth, although Proxima b, an Earth-size planet in a temperate orbit a mere 4.2 light-years away, whose discovery was announced last month, sounds as if it could be a decent place to live.

The discovery of exoplanets has been a source of special excitement to the scientists of the so-called SETI movement — the search for extraterrestrial intelligence. As the historian Lawrence Squeri notes in his engaging chronicle **WAITING FOR CONTACT: The Search for Extraterrestrial Intelligence (University Press of Florida, $26.95)**, the movement, which began in earnest in the 1960s, would have been "pointless" had there not turned out to be planets beyond our solar system. Where else would alien civilizations reside? Just as significant to SETI enthusiasts has been the very high percentage of stars that, when scrutinized, have revealed themselves to host a planet. The percentage is high enough that we can basically assume each of the 400 billion stars in our galaxy has a planetary companion. That's a lot of potential places for alien life to arise, evolve and — ideally using electromagnetic signals at a frequency we're monitoring — drop us a line.

SETI's principal methodology was first outlined by two physicists in the journal Nature in 1959. They proposed that the radio telescope, a new technology used to "see" objects in space thanks to the radio waves they emit, could also be used to send and receive interstellar messages with aliens at the speed of light. The next year, at an observatory in West Virginia, an astronomer named Frank Drake conducted the first such search for alien broadcasts, aiming a radio telescope at two stars about 10.5 and 12 light-years away. It was a crapshoot, and it was unsuccessful: Aliens had not, in fact, been sending us messages from those stars 10.5 or 12 years earlier — at least not at the frequency Drake guessed aliens would use. The next step for the movement, as Squeri writes, was to demonstrate this was "a cutting-edge experiment rather than a lost weekend for reputable scientists who had read too much science fiction."

The second American SETI search didn't take place until 1971, more than a decade later. (Securing time with a radio telescope is a competitive affair.) In the interim, and indeed throughout the movement's history, SETI scientists spent much of their time justifying and promoting their project. In this capacity, Squeri argues, SETI amounted to more than just a scientific enterprise; it was a kind of "offbeat political movement," a utopian ideology and perhaps surrogate religion — with aliens serving as our more enlightened counterparts. SETI arose amid widespread fears that nuclear weapons and population growth had made the human race a threat to itself. Many of the movement's founders were political progressives who believed that any alien civilization advanced enough to contact us would have survived as long as it did because it had solved such problems, presumably by mastering a system of world government. Drake, who later observed that many SETI supporters like himself were exposed in their youth to "fundamentalist religion," entertained hopes that alien wisdom would help humans end war and cure cancer.

Drake's other legacy to the movement was an equation he devised in 1961 for estimating the number of alien civilizations in our galaxy capable of communicating with us. The astrophysicist Neil deGrasse Tyson, in **WELCOME TO THE UNIVERSE: An Astrophysical Tour (Princeton University, $39.95),** revisits the Drake equation using contemporary data. The equation holds that the number of communicating alien civilizations is a function of seven variables, starting with the rate at which new stars are born in our galaxy, the fraction of these stars that host planets and the number of planets per star that are habitable. In 1961, scientists could fill in only one variable; the other six were sheer guesswork. With our advanced understanding of the cosmos, Tyson — whose book is written with the astrophysicists Michael A. Strauss and J. Richard Gott — is able to work out, in some technical detail, a more sophisticated estimate. The verdict? According to his calculations, we might expect to find as many as 100 alien civilizations in our galaxy communicating with radio waves right now. "So," he concludes, "we have a chance."

In **ALL THESE WORLDS ARE YOURS: The Scientific Search for Alien Life (Yale University, $30),** a lively introduction to the field of astrobiology, the astronomer Jon Willis concurs that exoplanets are an exciting development. But because our first contact with alien life, he suspects, "is likely to be a meeting of microbes rather than a meeting of minds," he is more optimistic about finding a simple organism elsewhere in our own solar system. Perhaps it will be an "extremophile" bacterium of the sort scientists have recently found on Earth living in conditions previously assumed to be fatal (like volcanic hot springs and subzero temperatures). Three of Willis's top contenders for habitats are Mars, Jupiter's moon Europa and Saturn's moon Titan. But if forced to choose a single project for both feasibility and promise, he would send a spacecraft on a flyby mission to Saturn's ice moon, Enceladus. The craft would collect (and bring back to Earth) icy particles and gases from Enceladus's subsurface liquid ocean, whose chilly waters are regularly spouted into space via geysers. Maybe we'd find something alive in that stuff.

Do planetary scientists, with their arcane interests and occasional searches for alien life, strike you as an exotic tribe? Do they seem to require decoding by an anthropologist who conducted fieldwork in their midst? If so, you might consult Lisa Messeri's study **PLACING OUTER SPACE: An Earthly Ethnography of Other Worlds (Duke**

University, paper, $23.95). Messeri, an anthropologist whose undergraduate degree is in aeronautical and astronautical engineering, spent 15 months working with — and closely observing the customs of — several groups of planetary scientists, including astronomers at a Chilean observatory, NASA researchers creating 3-D maps of Mars and exoplanet scientists at M.I.T. What she was trying to understand was their "pursuit of planetary place." By this phrase she means to identify a rarely considered aspect of their professional practice: how these scientists, whose objects of study exist at distances and scales and time-frames that can defy human grasp, nonetheless manage to conceive of these uncanny things more intimately as "places" or "worlds," thus allowing the scientists to better engage with them.

Exoplanet scientists offer the best example of what Messeri is talking about. Exoplanets are not visible from Earth, even with the assistance of the most powerful telescopes; instead, they are inferred from subtle changes in the light from the stars they orbit. A slight dimming in the light, a slight wobble in the light — it is from such fine measurements that astronomers determine an exoplanet's existence, the radius of its orbit, its mass and its density. From its density, they extrapolate what it is made of. In practice, this means exoplanet scientists pass their days looking not at planets but at data: light curves and radial velocity graphs and theoretical models. To become an exoplanet scientist, Messeri shows (in part by undergoing some training herself), is to learn to see and convey these abstractions as something more relatable — as "super-Earths" or "mini-Neptunes" or such. "To excite the community about a particular visualization," as Messeri nicely puts it, "is to convince them that the image contains a world." And to really excite the community, presumably, is to convince them that a world contains little green men.

JAMES RYERSON is a senior staff editor for The Times's Op-Ed page.

Flying Saucers and Other Fairy Tales

OPINION | BY ROSS DOUTHAT | DEC. 23, 2017

I AM COMPLETELY in favor of federal spending on U.F.O. research, an outlay whose existence was revealed to surprisingly little paranoid excitement by this newspaper last week. It is a sign of civilizational health to devote excess dollars to the scientific fringe, and to hope that bizarre secrets still await discovery even in our satellite-surveilled world. So good for Harry Reid and his little-green-men-obsessed billionaire pal for keeping the flame of weird curiosity alive.

But I also doubt that such research will ever prove that the strange lights and vessels filmed by human pilots actually belong to a starfaring species that's come to our planet to study, experiment and eventually offer us a hand up or else ruthlessly invade. Other sapient species may indeed be out there, but the most parsimonious explanation for all the U.F.O. encounters since Roswell is not that our nuclear testing or space program finally inspired the galaxy to come see what humanity is all about.

Rather, it's that our alien encounters, whether real or imaginary, are the same kind of thing as the fairy encounters of the human past — part of an enduring phenomenon whose interpretations shift but whose essentials are consistent, featuring the same abductions and flying crafts and lights and tricks with crops and animals and time and space, the same shape-shifting humanoids and sexual experiments and dangerous gifts and mysterious intentions.

This was the argument of Jacques Vallée, a French-born scientist and a wonderful character in the annals of ufology, who wrote a wild book in the heady year of 1969 called "Passport to Magonia: From Folklore to Flying Saucers," which The Times's U.F.O.-spending scoop gave me an excuse to read.

Vallée's conclusion is basically the reverse of Erich von Däniken's thesis in "Chariots of the Gods," published to better sales the prior year. Where von Däniken argued that old myths and biblical tales alike contain evidence of ancient alien visitations (an idea picked up, most recently, by Ridley Scott's "Alien" prequels), Vallée suggested that contemporary U.F.O. narratives are of piece with stories about Northern European fairies and their worldwide kith and kin — and that it's more reasonable to think that we're reading our space age preoccupations into a persistent phenomenon that might be much weirder than a simple visitation from the stars.

This quasi-magical thesis made Vallée, as he put it, a "heretic among heretics" — the U.F.O. believer who rejected the U.F.O. community's hope that their efforts could one day be incorporated into the normal sciences and lead us to some Spielbergian first contact. But his arguments for the basic continuity between folklore and flying saucers are quite compelling, and I suspect he's correct about the commonality of these experiences …

… Which is not, of course, to say that they reflect the genuine existence of some fifth-dimensional fairyland, from whence morally ambiguous beings emerge to play tricks upon our race. Certainly for most sensible secular scientific-minded people, to say that our era's close encounters are of the same type as encounters with the unseelie court of faerie is to say that they are all equally imaginary, proceeding from internalized fancies and hallucinatory substances and late-night wrong turns, plus some common evolved subconscious that fears shape-shifting tricksters in modern Nevada no less than in the mists around Ben Bulben.

But if this rationalist assumption seems natural these days, it is not necessarily permanent. The educated class of Victorian England went wild for fairies and spirits in the heyday of scientific optimism, and both Vallée and von Däniken offered up their books amid the Age of Aquarius's similar craze. (Just read Sally Quinn's tales of murderous hexes in her recent memoir to recall how old-fashioned in their magical thinking the New Age's devotees could become.)

Sometimes our own elite opinion seems to be shopping for a new religion: I have read books in the last year pitching versions of Buddhism, pantheism and paganism to the post-Christian educated set. For such shoppers, the striking overlap between U.F.O.s and fairy stories might eventually become an advertisement for an updated spiritualist cosmology, not a strike against it — especially if woven together with multiverse and universe-as-simulation hypotheses that imply a kind of metaphysics of caprice.

Meanwhile those of us who remain Christian — and yes, this is a Christmas column, U.F.O.s and all — can be agnostic about all these strange stories, not reflexively dismissive, since Christianity does not require that all paranormal experiences be either divinely sent or demonic or imaginary.

Rather the Christian idea is that whatever capricious powers may exist, when the true God enters his creation, he does so honestly, straightforwardly, in a vulnerable and fully human form — and exposes himself publicly, whether in a crowded stable or on an execution hill. So the glamour of U.F.O.s, like the glamour of faerie, is an understandable object of curiosity but a dangerous object for any kind of faith. The only kind of God worth trusting is the kind who does not play tricks.

ROSS DOUTHAT is an Op-Ed columnist for The New York Times.

U.F.O.s: Is This All There Is?

COLUMN | BY DENNIS OVERBYE | DEC. 29, 2017

Hey, Mr. Spaceman,
Won't you please take me along?
I won't do anything wrong.
Hey, Mr. Spaceman,
Won't you please take me along for a ride?

SO SANG THE BYRDS in 1966, after strange radio bursts from distant galaxies called quasars had excited people about the possibility of extraterrestrial intelligence.

I recalled those words recently when reading the account of a pair of Navy pilots who were outmaneuvered and outrun by a U.F.O. off the coast of San Diego back in 2004. Cmdr. David Fravor said later that he had no idea what he had seen.

"But," he added, "I want to fly one."

His story was part of a bundle of material released recently about a supersecret $22 million Pentagon project called the Advanced Aerospace Threat Identification Program, aimed at investigating U.F.O.s. The project was officially killed in 2012, but now it's being resurrected as a nonprofit organization.

Disgruntled that the government wasn't taking the possibility of alien visitors seriously, a group of former defense officials, aerospace engineers and other space fans have set up their own group, To the Stars Academy of Arts & Science. One of its founders is Tom DeLonge, a former punk musician, record producer and entrepreneur, who is also the head of the group's entertainment division.

For a minimum of $200, you can join and help finance their research into how U.F.O.s do whatever it is they do, as well as telepathy and "a point-to-point transportation craft that will erase the current travel limits of distance and time" by using a drive that "alters the space-time metric" — that is, a warp drive going faster than the speed of light, Einstein's old cosmic speed limit.

"We believe there are transformative discoveries within our reach that will revolutionize the human experience, but they can only be accomplished through the unrestricted support of breakthrough research, discovery and innovation," says the group's website.

I'm not holding my breath waiting for progress on telepathy or warp drive, but I agree with at least one thing that one official with the group said. That was Steve Justice, a former engineer at Lockheed Martin's famous Skunk Works, where advanced aircraft like the SR-71 high-altitude super-fast spy plane were designed.

"How dare we think that the physics we have today is all that there is," he said in an interview published recently in HuffPost.

I could hardly agree more, having spent my professional life in the company of physicists and astronomers trying to poke out of the cocoon of present knowledge into the unknown, to overturn Einstein and what passes for contemporary science. Lately, they haven't gotten anywhere.

The last time physicists had to deal with faster-than-light travel was six years ago, when a group of Italy-based physicists announced that they had seen the subatomic particles known as neutrinos going faster than light. It turned out they had wired up their equipment wrong.

So far Einstein is still the champ. But surely there is so much more to learn. A lot of surprises lie ahead, but many of the most popular ideas on how to transcend Einstein and his peers are on the verge of being ruled out. Transforming science is harder than it looks.

While there is a lot we don't know, there is also a lot we do know. We know how to turn on our computers and let gadgets in our pocket navigate the world. We know that when physical objects zig and zag through a medium like air, as U.F.O.s are said to do, they produce turbulence and shock waves. NASA engineers predicted to the minute when the Cassini spacecraft would dwindle to a wisp of smoke in Saturn's atmosphere last fall.

DEPARTMENT OF DEFENSE

A U.F.O. spotted by Navy pilots near San Diego in 2004.

In moments like this, I take comfort in what the great Russian physicist and cosmologist Yakov Zeldovich, one of the fathers of the Soviet hydrogen bomb, once told me. "What science has already taken, it will not give back," he said.

Scientists are not the killjoys in all this.

In the astronomical world, the border between science fact and science fiction can be very permeable, perhaps because many scientists grew up reading science fiction. And astronomers forever have their noses pressed up against the window of the unknown. They want to believe more than anybody, and I count myself among them.

But they are also trained to look at nature with ruthless rigor and skepticism. For astronomers, the biggest problem with E.T. is not the occasional claim of a mysterious light in the sky, but the fact that we are not constantly overwhelmed with them.

Half a century ago, the legendary physicist Enrico Fermi concluded from a simple back-of-the-envelope calculation that even without warp drive, a single civilization could visit and colonize all the planets in the galaxy in a fraction of the 10-billion-year age of the Milky Way.

"Where are they?" he asked.

Proponents of SETI, the search for extraterrestrial intelligence, have been debating ever since. One answer I like is the "zoo hypothesis," according to which we have been placed off-limits, a cosmic wildlife refuge.

Another answer came from Jill Tarter, formerly the director of research at the SETI Institute in Mountain View, Calif. "We haven't looked hard enough," she said when I asked her recently.

If there was an iPhone sitting under a rock on the Moon or Mars, for example, we would not have found it yet. Our own latest ideas for interstellar exploration involve launching probes the size of postage stamps to Alpha Centauri.

In the next generation, they might be the size of mosquitoes. By contrast, the dreams of some U.F.O. enthusiasts are stuck in 1950s technology.

Still, we keep trying.

Last fall when a strange object — an interstellar asteroid now named Oumuamua — was found cruising through the solar system, astronomers' thoughts raced to the Arthur C. Clarke novel "Rendezvous With Rama," in which the object was an alien spaceship. Two groups have been monitoring Oumuamua for alien radio signals, so far to no avail.

Meanwhile, some astronomers have speculated that the erratic dimming of a star known as "Boyajian's star" or "Tabby's star," after the astronomer Tabetha Boyajian, could be caused by some gigantic construction project orbiting the star. So far that has not worked out, but none of the other explanations — dust or a fleet of comets — have, either.

A pair of Harvard astronomers suggested last spring that mysterious sporadic flashes of energy known as fast radio bursts coming from far far away are alien transmitters powering interstellar spacecraft carrying light sails. "Science isn't a matter of belief, it's a matter of evidence," the astronomer Avi Loeb said in a news release from Harvard. "Deciding what's likely ahead of time limits the possibilities. It's worth putting ideas out there and letting the data be the judge."

U.F.O. investigations are nothing new. The most famous was the Air Force's Project Blue Book, which ran from 1952 to 1970 and examined more than 12,000 sightings.

Most U.F.O. sightings turn out to be swamp gas and other atmospheric anomalies, Venus, weird reflections or just plain hoaxes. But there is a stubborn residue, a few percent that resist easy explication, including now Commander Fravor's story. But that's a far cry from proving they are alien or interstellar.

I don't know what to think about these stories, often told by sober, respected and professional observers — police officers, pilots, military officials — in indelible detail. I always wish I could have been there to see it for myself.

Then I wonder how much good it would do to see it anyway.

Recently I ran into my friend Mark Mitton, a professional magician, in a restaurant. He came over to the table and started doing tricks. At one point he fanned the card deck, asked my daughter to pick one, and then asked her to shuffle the deck, which she did expertly.

Mr. Mitton grabbed the deck and sprayed the cards in the air. There was my daughter's card stuck to a mirror about five feet away. How did it get there? Not by any new physics. Seeing didn't really help.

As modern psychology and neuroscience have established, the senses are an unreliable portal to reality, whatever that is.

Something might be happening, but we don't know what it is. E.T., if you're reading this, I'm still waiting to take my ride.

DENNIS OVERBYE writes the Out There column for The New York Times.

They've 'Seen Things'

BY ROZETTE RAGO | AUG. 14, 2018

A group in Los Angeles has attracted U.F.O. enthusiasts from all over the world. They've formed together around the common question: What are these things in the sky, exactly, and how can they know more about them?

LOS ANGELES — Robert Bingham has "seen things." When he was 39, he looked skyward and noticed a worm-shaped ship about 20 feet tall zipping through the clouds.

Unusual things kept popping up around him — or above him, rather. He saw a saucer and some flying objects shaped like beans next. He snapped a picture.

For over ten years, he kept his sightings to himself. That changed in 2010, when his neighbor came over to do some plumbing work. Mr. Bingham showed him his photos. The neighbor asked if he could invite his brother, who was very interested in unidentified flying objects, or U.F.O.s.

In awe of what they saw, they asked if they could invite more people to speak with Mr. Bingham — 40 more, actually. More than eight years ago, that was the first meeting of what is now known as "Summon Events with Robert Bingham," at a park in Los Angeles across the street from where Mr. Bingham worked as a security guard.

Mr. Bingham, 62, an unassuming man who describes himself as shy, has become the nexus of a community of U.F.O. hunters in Los Angeles, fervent believers who come together to share their stories and persuade skeptics that extraterrestrial communications aren't just a conceit for television shows.

Since then, he has attracted U.F.O. enthusiasts from all over the world, drawn together by the same questions: What are these things in the sky, exactly, and how can we learn more about them?

While there is just not enough documentation or scientific evidence to begin to explain or even confirm these sightings, that doesn't stop

From left, Yasmin Joyner, Jim Martin, John Graf, Fausto Perez and Mallory Jackson at a group summoning at Ms. Joyner's home in Los Angeles.

the dozens of people that once a year descend on the same park to watch and assist Mr. Bingham as he tries to summon the "objects," as they call them, and also to hang out with other enthusiasts who have turned into friends.

"It's a great community because you can talk about anything and you're not worried about being called crazy," said Hans Boysen, 53, who has participated in the last seven summoning sessions with Bingham since 2011.

Other groups, like the U.F.O. and Paranormal Research Society, don't organize sighting sessions, but rather focus on discussions, often with speakers who talk about their research and experiences. A nonprofit group called the Mutual U.F.O. Network, or Mufon, founded in 1969, has over 4,000 members worldwide and convenes a yearly symposium. This year, a former Pentagon intelligence officer, Luis Elizondo, will give the keynote address.

ROZETTE RAGO FOR THE NEW YORK TIMES

Robert Bingham delivered the opening remarks at his annual sighting event in April.

Last year, The New York Times conducted interviews and obtained records pertaining to the $22 million spent on the Advanced Aerospace Threat Identification Program. The program — parts of which remain classified — investigated reports of unidentified flying objects, according to Defense Department officials. According to the article, officials insisted that the effort had ended after five years, in 2012. The article also stated that Sara Seager, an astrophysicist at M.I.T., cautioned that not knowing the origin of an object does not mean that it is from another planet or galaxy. "When people claim to observe truly unusual phenomena, sometimes it's worth investigating seriously," she said. But, she added, "what people sometimes don't get about science is that we often have phenomena that remain unexplained."

As much as scientists deal with probabilities, they rely on data and the reality is, no matter how many videos people upload on YouTube, they're simply not enough to draw any definitive conclusions from.

ROZETTE RAGO FOR THE NEW YORK TIMES

Ms. Joyner, left, received an inflatable alien as a birthday present from Mr. Perez and Trinity.

But that doesn't stop this community from searching. Many in the community that forms around Mr. Bingham believe that the multimillion-dollar alien research efforts of the former Blink-182 guitarist and singer Tom DeLonge are just the beginning to finding out some answers. Mr. DeLonge made headlines after To the Stars Academy of Arts & Science, a research group he founded — Mr. Elizondo is its director of global security and special programs — released declassified footage from the Department of Defense and continues his efforts.

Angel Llewellyn, 49, drove to the event from San Jose, Calif., for a second year as a form of pilgrimage. She said she started seeing things right after attending Mr. Bingham's event for the first time.

"It's like he charges you," she said. "He teaches you how to call them and what to think and they just, boom, boom, boom. It's like fishing. You never know what you're going to get."

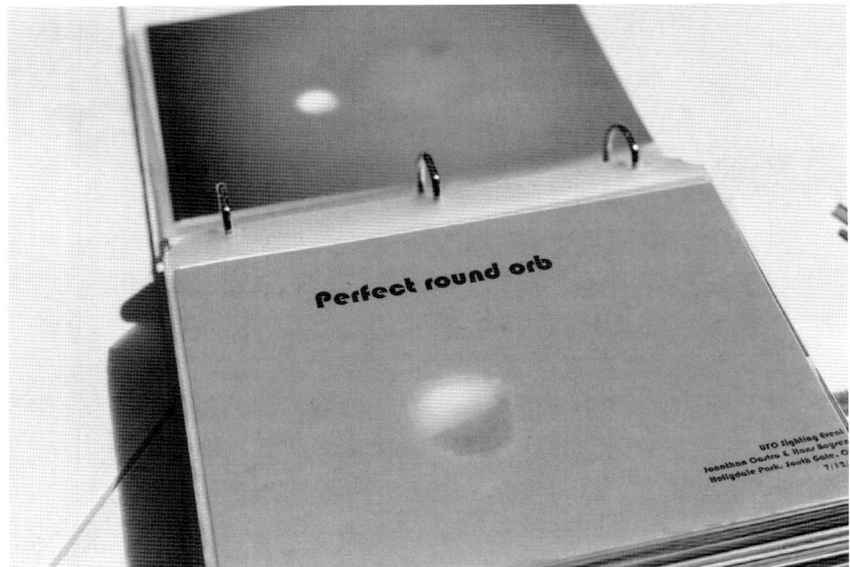

ROZETTE RAGO FOR THE NEW YORK TIMES

A page from Mr. Boysen's book of printed photographs from various U.F.O. sightings.

That's not to say that everyone in the group is on the same page. There's a variety of ideologies — some beliefs are grounded in religion, while some prefer a more scientific foundation — and different levels of intensity. Rafael Cebrian, 29, a two-time attendee, brought a curious friend who was visiting from Spain. He said it was about being open and being in the right state of mind. He went to Mr. Bingham's event last year on a friend's recommendation. He said he didn't think he was a skeptic anymore, but he also insisted he didn't know anything for sure.

Mr. Bingham may have brought them all together, but now factions abound.

One such group is the L.A. U.F.O. Channel, a monthly Meetup group founded by Mr. Perez and Hans Boysen, 53.

"I know they're never going to believe a video that they think was created in somebody's basement," Mr. Perez said. "I see the events as a bigger way to change people's minds."

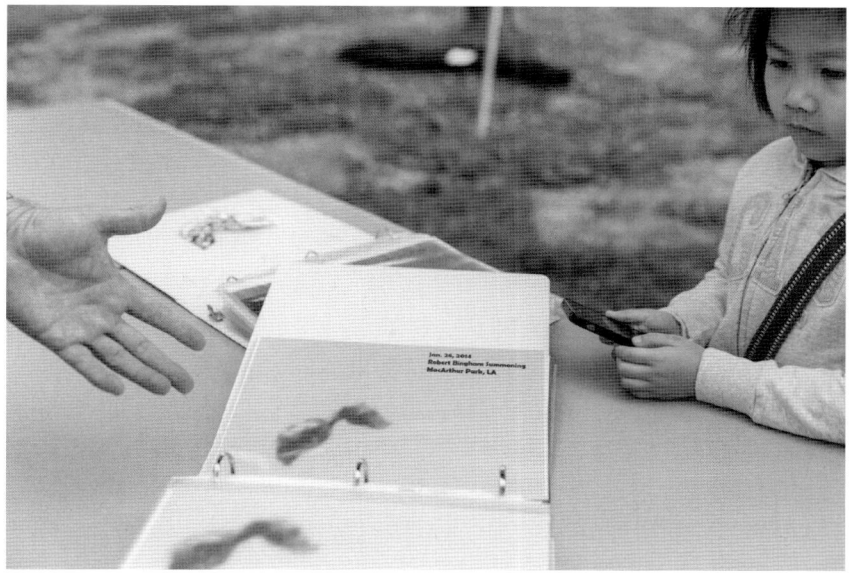

ROZETTE RAGO FOR THE NEW YORK TIMES

Mr. Boysen showed off images from previous sightings.

Mr. Boysen was once a skeptic himself and tried to contact Mr. Bingham in 2011. "Like most people, I suspect, I thought he was a nut case or lunatic," he said. But as a skeptic, he was "willing to keep an open mind and look at what he had to offer." He instead found Mr. Perez, who helped him discover his passion for the unknown flying objects in the sky, something he has always been interested in since he was a child.

Yasmin Joyner, 35, an artist, says she has a more straightforward approach to sightings and doesn't like to engage in conspiracy theories. "Unidentified flying objects: That's what a U.F.O. is, right?" she said. "I'm not saying it's an alien. I'm not saying it's from another planet. I'm not saying it's even a being. I don't know!"

"I try to go with what I can say I know," she continued. "I'm not going to look at something and try to equate that it might be an animal or something biological. I don't know that and I will never claim to know that." She recently formed another group, Indigo Army, that she hopes will attract a younger and more diverse crowd. Because it's still

ROZETTE RAGO FOR THE NEW YORK TIMES

From left, Ms. Joyner, Mr. Graf, Mr. Martin, Ms. Jackson and Mr. Perez concentrating for a group summoning at Ms. Joyner's house.

a small group, its members have been able to organize sightings at one another's houses and nearby parks.

For Ms. Joyner, going public with her belief of U.F.O.s and extraterrestrial communications wasn't a decision she took lightly. She understands that there are consequences to being outspoken about beliefs that many people may deem weird or crazy.

"I think my family was a bit worried that I had snapped or something, but once they saw my footage and what I was seeing, they understood," she said. Her mother is fully supportive and claims to have seen unknown objects flying in the sky near her home in Los Angeles multiple times.

Mallory Jackson, 26, who attends Meetup hangouts led by Mr. Bingham and the Indigo Army, says she finds it difficult to maintain relationships outside the U.F.O. community with people who might not be as understanding of her interests. She discovered Mr. Bingham's

An attendee hugs Mr. Bingham at the April sighting event.

event through a friend she met at a metaphysical center that does reiki healing, with whom she later confided in about her sightings. She said she became friends with several members "right away" before later meeting Jim Martin, 38, another longtime attendee of Mr. Bingham's events. Mr. Martin is now her boyfriend.

"When we do our events, you'll see all ages, all ethnicities, all genders," Ms. Jackson said. "It's beautiful, and we're all just trying to figure it out as we go. We don't know what they are, but we all make our assumptions and best guesses."

Around noon on a punishingly hot day in April, Mr. Bingham gathered everyone around him for the first so-called group summoning of the day. "Let's make this world a better place," he said. "Enjoy this day because it's going to be incredible." As he concluded his opening remarks, he turned around and led the group to hope for something good to show up.

So, how exactly does the group try to summon U. F. O.s? Everyone has a different method, but most agree that it's similar to meditating. Some say that they feel physical sensations when they do it. The most important thing, Ms. Joyner said, is to focus. At the event in April, some participants closed their eyes and stood silently. Some stared intently into the sky. A few newcomers simply looked around, appearing confused.

As soon as someone in the group spots something, they yell at Mr. Boysen, who has a telescope connected to a camera and a screen that shows what he's seeing. Once he spots it, he holds out his arm to ask for someone to guide him back to his chair without losing sight of the object. He adjusts his telescope in search of it, while the guide looks at the screen to tell him if he's got it or not. Everything is recorded as video footage that he will later stabilize using Adobe After Effects, a video-editing software. "Nothing more," Charles Cassey, 50, Mr. Boysen's frequent guide, insists. "All he does is stabilize the footage so it's not so shaky." Mr. Cassey puts his hand on the focus knob and adjusts it as soon as Mr. Boysen centers on the object.

Effectively capturing objects in the sky from dozens of miles away requires a significant financial investment. Ms. Joyner uses a camera with a lens that can magnify objects up to 40 times, which could capture things in the sky as if they were only 20 feet away, but in broad daylight, it can still get a little tricky. Mr. Martin works his way around such issues by combining a super-zoom lens with infrared, which helps him spot things in the sky easier. He is frequently praised by people in the community for his consistently high-quality footage, and his YouTube page is full of well-edited clips from various sightings, as well as other U.F.O.-related videos.

The group's members encounter their fair share of people who don't believe them online. It's not unusual to see U.F.O. believers try to debunk each other's videos by dissecting them frame by frame. "Some people's opinions are so hardened that I just let them think what they want and focus on research that I think is valid," Mr. Martin said.

"Through the videos that I post, I try and set an example of what I think is good evidence to kind of rise against some of the misidentification." This forms a core part of the new mission of the Indigo Army: to look harder at the evidence and be more selective about what the group claims to be a U.F.O.

Ms. Joyner agrees, and thinks they can only get better by scrutinizing everything closely, including their own work. "There are videos that I have that I questioned a year later and I don't have any issues doing that," she said. "If we have a problem with being wrong, there's never going to be any truth."

"Oh, that video? Might've been just birds. And I'm O.K. with that. But this video? This video isn't birds."

CHAPTER 2

Aliens and the U.S. Government

For those interested in the possibility of aliens and extraterrestrial life, the United States government has long been a source of fascination and intrigue. From reports of Area 51, a classified Air Force base in a remote area of Nevada that was rumored to host crashed U.F.O.s, to secret projects designed to search for life on other planets, the U.S. government is invested in the hunt for aliens. The articles in this chapter show how the U.S. government sees its role as an arbiter between life on Earth and possible life in outer space.

Visitors From Outer Space, Real or Not, Are Focus of Discussion in Washington

BY ANDREW SIDDONS | MAY 3, 2013

WASHINGTON — While President Obama was promoting an immigration overhaul in Mexico, six former members of Congress gathered two blocks from the White House to consider what they see as the enforced government secrecy surrounding another kind of visitor: the kind who come from a lot farther away.

Every day this week, the former legislators presided over panels made up of academics and — former, of course — government and military officials, who were there to discuss their research or their own eyewitness accounts of unidentified flying objects and the extraterrestrials who presumably would have occupied them.

"Something is monitoring the planet, and they are monitoring it very cautiously, because we are a very warlike planet," said Mike Gravel, a former Democratic senator from Alaska who ran in both the Democratic and Libertarian presidential primaries in 2008.

Mr. Gravel and his fellow panelists were assembled by the Paradigm Research Group, which says it is committed to ending the government's "truth embargo" on the existence of extraterrestrial life. The lawmakers were there in hopes that their presence and political credibility would be enough to persuade Congress to take the issue seriously.

"I've been exploring how we might get this issue out of the shadows of the lunatic fringe," said Roscoe G. Bartlett, a former Republican representative from Maryland. Before his defeat last year, Mr. Bartlett was known for sounding the alarm on the threat posed to the nation's energy infrastructure by electromagnetic pulse, or EMP, the shock wave from a nuclear weapon detonated beyond the earth's atmosphere.

Called the Citizen Hearing on Disclosure, the event might have been mistaken as advocacy for government transparency, and some of the panelists had impressive résumés.

"I've come to understand and appreciate the importance of open, transparent government and the power of truth," said Paul T. Hellyer, who served as Canadian minister of defense during the 1960s.

"We are not alone in the cosmos," he added.

One reason the ex-members of Congress agreed to sit on the dais and ask questions may have been curiosity.

"Our country has trivialized it, has made it a joke, has made it green people with horns sticking out," said Carolyn Kilpatrick, a Democratic representative from Michigan who lost her seat in 2010. "Now I find that it's much more than that. And it's not a joke. And there is scientific data that there may be something there."

Another reason might have been the $20,000 the organizers said they paid each panelist. But they are still maintaining a healthy skepticism.

"Just because the government might have had a document about how to handle extraterrestrials doesn't mean there were any," said Merrill Cook, a Republican from Utah who was twice elected to the House.

The panels this week have been low-hanging fruit for the news media while President Obama is out of town and Congress is out of session, and not all of the people who study U.F.O.'s think the meetings will help them improve their stature in Washington.

"There really is something to this issue, and there is a serious side to it, but that's not what's being presented as this event," said Leslie Kean, a journalist and author of "U.F.O.'s: Generals, Pilots and Government Officials Go on the Record," a collection of firsthand accounts by people who believe they saw them.

The conclusion that U.F.O.'s are proof of extraterrestrial life is misguided, she said, and the people who broadcast that belief hindered support for real scientific research.

Despite the ridicule that usually accompanies the discussion of U.F.O.'s, they have been quietly talked about in corridors of power here. Some panelists at the event this week counted among true believers John D. Podesta, a chief of staff in President Bill Clinton's White House, because of his role in Executive Order 12958, which requires the declassification of most government documents over 25 years old.

But the possible existence of extraterrestrial life is not exactly why he believes in government transparency, Mr. Podesta said.

"At the end of the day, there are going to be people who say that even if you did that, there must be other files that exist that you're not disclosing," he said in an interview.

But objects in the sky have piqued his interest. In June 2011, the Center for American Progress hosted government officials, from the Pentagon, NASA and the Department of Transportation, as well as

Congressional staff and former officials from intelligence organizations, for a briefing by Ms. Kean and experts from academia and foreign militaries.

The private briefing was organized to discuss a proposal that the government establish a small office of two staff members who would selectively investigate mysterious skyward sightings and seek to understand them by applying scientific method. The proposal did not refer to U.F.O.'s, but rather, U.A.P.'s, unidentified aerial phenomena, as if those who drew up the proposal were keenly aware of how their objective could be perceived.

"They were interesting, credible people who had observed aerial phenomena that were unexplained and worthy of additional follow-up," Mr. Podesta said. "Going back and looking at and declassifying whatever government documents exist is a smart thing to do."

C.I.A. Acknowledges Area 51 Exists, but What About Those Little Green Men?

BY ADAM NAGOURNEY | AUG. 22, 2013

RACHEL, NEV. — The Little A'Le'Inn has been an unlikely tourist destination in the Mojave Desert for nearly 25 years, selling souvenirs — from green alien coffee cups to E.T. Highway T-shirts — dedicated to the notion that we are not alone. Understandably.

Nine miles up a nearby dirt road is the top-secret military installation known as Area 51, whose murky provenance fueled decades of speculation about extraterrestrial aliens and kept the U.F.O.-hunting tourists coming.

Or rather, the top-secret military installation not known as Area 51 — at least until last week, when the C.I.A. released a classified report on the history of the U-2 spy plane, which officially acknowledged what everyone here has long known: There is a secret military testing base at Groom Lake called Area 51. It is 150 miles north by car from Las Vegas, in a vast expanse of utterly empty scabland, desert and mountain, and signs reading "No gas station next 150 miles."

The report, released after eight years of prodding by a George Washington University archivist researching the history of the U-2, made no mention of colonies of alien life, suggesting that the secret base was dedicated to the relatively more mundane task of testing spy planes.

But no matter. Even this little bit of validation was welcome in Rachel, which claimed a population of 57 as of last Tuesday afternoon, and where the tourists have not been coming at quite the pace they once did. The movies and television shows that once fed an international fixation with aliens secreted at Area 51 — from "The X Files" to movies like "Independence Day" and "Paul" — are, with the passage of time and the inevitable rise of new subjects of national interest, not quite as gripping as they once were.

"We have a guest book, but it's gone by the wayside. Shelby, do you know where our guest book has disappeared to?" Pat Travis, 70, the owner of the Little A'Le'Inn, asked a waitress behind the bar.

"It's really strange to not have it out for all of our customers to sign," Ms. Travis said with a sigh. "Would you bring it to me?"

Ms. Travis — who recounted being awakened one night by a bright light shot from a U.F.O. that came through the center of the back door — said she expected the C.I.A. acknowledgment to bring a rush of customers through her doors.

They will want to know how to find Area 51 and how to spot a U.F.O. in the pitch-black night skies here, and they will shop from the shelves lined with green alien shot glasses, coffee cups and guitar picks and even Area 51 Wine (produced for the inn by a winery in Northern California).

ISAAC BREKKEN FOR THE NEW YORK TIMES

A souvenir shop in Hiko, Nev., not far from Area 51, a longtime locus of conspiracy theories.

"Every time there is another story out, people come out," she said. "They want to know how to get to that area. Where it is. The more there is, the more you talk about it, the more it goes on and on."

Rachel was fairly deserted the other day, save for a family from Seattle on a drive from Las Vegas to Yosemite that made a U.F.O. detour at the urging of their 16-year-old, Hank Reavis. His arms full of Area 51 T-shirts as his father reached for his wallet, Hank said he wanted to see for himself the place featured in movies like "Paul."

Asked if they would visit Area 51 itself, Hank's father, Gil, a retired logger, answered "Yes." Hank corrected him.

"We won't get in there, Dad," he said.

That observation was confirmed after a nine-mile drive up Back Gate Road to a back entrance of the base. Or at least, one assumes it was the back entrance to the base, given the six separate WARNING! signs prohibiting picture-taking or going beyond the two guard gates with flashing red lights.

"If you pass the gate, they can shoot you, I think," said Niklas Gartler, 17, of Vienna, who came here with his uncle from Los Angeles.

Well maybe not shoot; the signs promise six months in prison for trespassers. The greater threat, in truth, might be the rattlesnakes that infest the roads and trails here during the hot summer months.

The report, "The Central Intelligence Agency and Overhead Reconnaissance: The U-2 and Oxcart Programs, 1954-1974," was released, albeit in a redacted form, at the request of Jeffrey T. Richelson, a senior fellow with the National Security Archive at George Washington University. "There certainly was — as you would expect — no discussion of little green men here," Mr. Richelson said. "This is a history of the U-2. The only overlap is the discussion of the U-2 flights and U.F.O. sightings, the fact that you had these high-flying aircraft in the air being the cause of some of the sightings."

Mr. Richelson said he was not looking for information on Area 51. "That was sort of a bonus," he said.

No one here seems to take themselves too seriously. The prevailing attitude is reflected in the name of the restaurant, The Little A'Le'Inn (say it aloud). It sits right off Extraterrestrial Highway, as they call State Route 375. There is an "Earthlings welcome" sign above the parking lot.

But everyone seems assured that aliens are here, that U.F.O.s are dancing through the desert skies, and that the government has never been straight about what it was up to.

"I never had any doubt," said Pam Kinsey, a housekeeper here. "I watch the lights every morning. I get up at 4:30 to send my kid to school. I know they are there."

Hank's mother, Sally, said she was keeping an open mind.

"But how can we be the only ones?" she asked. "I'll tell you this, they certainly picked a beautiful state to come to. They couldn't have done much better than Nevada."

Howard Baral, a Los Angeles entertainment accountant and Niklas's uncle, said that he made the trip out here — it is a three-hour drive from Las Vegas — to make his nephew happy.

"Since he was a little kid, he has always been enthralled with alien lore and Area 51," Mr. Baral said. "His dream was to visit it. It wasn't my first choice."

That said, Mr. Baral said he was inclined to believe that there were aliens out there. "It's the middle of nowhere," he said. "What's the Air Force doing in the middle of nowhere?"

Annie Jacobson, the author of a book on the history of the area, said she doubted the acknowledgment would dampen interest in what lies behind the fences.

"It will only make people more curious, ask more questions," she said.

All of which is why Ms. Travis thinks it is time to get that guest book on display at the entrance of her little box of a restaurant.

"You are going to find people coming in here from different country, different places," she said, thumbing through pages of signatures from the past. "This needs to get back out. We need to get our little table out."

Hillary Clinton Gives U.F.O. Buffs Hope She Will Open the X-Files

BY AMY CHOZICK | MAY 10, 2016

WHEN JIMMY KIMMEL asked Hillary Clinton in a late-night TV interview about U.F.O.s, she quickly corrected his terminology.

"You know, there's a new name," Mrs. Clinton said in the March appearance. "It's unexplained aerial phenomenon," she said. "U.A.P. That's the latest nomenclature."

Known for her grasp of policy, Mrs. Clinton has spoken at length in her presidential campaign on topics as diverse as Alzheimer's research and military tensions in the South China Sea. But it is her unusual knowledge about extraterrestrials that has struck a small but committed cohort of voters.

Mrs. Clinton has vowed that barring any threats to national security, she would open up government files on the subject, a shift from President Obama, who typically dismisses the topic as a joke. Her position has elated U.F.O. enthusiasts, who have declared Mrs. Clinton the first "E.T. candidate."

"Hillary has embraced this issue with an absolutely unprecedented level of interest in American politics," said Joseph G. Buchman, who has spent decades calling for government transparency about extraterrestrials.

Mrs. Clinton, a cautious candidate who often bemoans being the subject of Republican conspiracy theories, has shown surprising ease plunging into the discussion of the possibility of extraterrestrial beings.

She has said in recent interviews that as president she would release information about Area 51, the remote Air Force base in Nevada believed by some to be a secret hub where the government stores classified information about aliens and U.F.O.s.

In a radio interview last month, she said, "I want to open the files as much as we can." Asked if she believed in U.F.O.s, Mrs. Clinton said: "I don't know. I want to see what the information shows." But she added, "There's enough stories out there that I don't think everybody is just sitting in their kitchen making them up."

When asked about extraterrestrials in an interview with The Conway Daily Sun in New Hampshire last year, Mrs. Clinton promised to "get to the bottom of it."

"I think we may have been" visited already, she said in the interview. "We don't know for sure."

While Americans typically point to issues like the economy and terrorism as top priorities for the next president, a desire for answers about aliens has inspired a passionate bloc of voters, who make their voices heard on social media.

Stephen Bassett, who lobbies the government on extraterrestrial issues, views a Clinton presidency as a chance to finally get the United States to disclose all it knows about life beyond Earth. Since November 2014, Mr. Bassett's organization has sent roughly 2.5 million Twitter messages to presidential candidates, elected officials and the news media urging a serious discussion of the issue.

"That was a storm, and now it's like a steady drip," Mr. Bassett said.

The movement viewed Mrs. Clinton's decision to correct Mr. Kimmel's use of the term U.F.O., which some view as loaded and rooted more in science fiction than in science, as a breakthrough because it "suggested she'd been briefed by someone and is not just being flippant," Mr. Buchman said.

In fact, Mrs. Clinton had been briefed. She was prepped by her campaign chairman, John D. Podesta, who is not only a well-respected Washington hand, having served as a top adviser to Mr. Obama and President Bill Clinton, but also a crusader for the disclosure of government information on unexplained phenomena that could prove the existence of intelligent life outside Earth.

"The time to pull back the curtain on the topic is long overdue," Mr. Podesta wrote in his foreword for the 2010 book "UFOs: Generals,

Pilots and Government Officials Go on the Record," by Leslie Kean, an investigative journalist.

Mrs. Clinton's position is not a political response to public sentiment — 63 percent of Americans do not believe in U.F.O.s, according to an Associated Press poll. But it reflects the decades of overlap between the rise to power of Bill and Hillary Clinton and popular culture's obsession with the universe's most mysterious questions.

In 1996, Mrs. Clinton was ridiculed after Bob Woodward reported, in his book "The Choice," that as first lady she had held discussions with her deceased role models, Eleanor Roosevelt and Mohandas K. Gandhi. The tabloid Weekly World News dreamed up sensational headlines about Mrs. Clinton's adopting an alien baby and having a "U.F.O. love nest."

The Clinton presidency also coincided with the hit television series "The X-Files" and movies like "Independence Day," which gave way to an era of fascination with the existence of aliens and the possibility of a government cover-up.

Mr. Podesta, an "X-Files" fanatic, ran a fan club for the show in the Clinton White House. "The 'X-Files' fan club would like to invite you and Mulder to lunch at the White House. Don't let the boss know," he wrote in a 1998 email, referring to the show's fictional F.B.I. agent Fox Mulder, according to White House documents. In 1999, Mr. Podesta had an "X-Files"-themed 50th birthday party that the Clintons attended.

When Mr. Podesta left the White House last year, he posted on Twitter: "Finally, my biggest failure of 2014: Once again not securing the #disclosure of the U.F.O. files. #thetruthisstilloutthere."

Mr. Podesta declined to comment for this article.

Mrs. Clinton, who speaks frequently about her childhood aspirations to be a NASA astronaut, has been sympathetic to Mr. Podesta's efforts.

In 1995, when she was photographed visiting Laurance S. Rockefeller, a billionaire philanthropist, in Jackson Hole, Wyo., she had tucked under her arm a copy of "Are We Alone?: Philosophical Implications of the Discovery of Extraterrestrial Life," by Paul Davies.

Before that meeting, John H. Gibbons, the former director of the White House Office's of Science and Technology Policy, had warned Mrs. Clinton about Mr. Rockefeller, who had spent years pressuring the government to release files relating to a 1947 crash near Roswell, N.M., that had become the source of theories about a cover-up of an alien spaceship.

He will "want to talk to you about his interest in extrasensory perception, paranormal phenomena and U.F.O.s," Mr. Gibbons wrote.

The meeting enthralled conspiracy theorists and, in turn, inspired Hollywood writers.

"If you look at our mythology, there are elements of those kinds of meetings," Chris Carter, the creator and executive producer of "The X-Files," said in an interview. Mr. Carter, who is supporting Mrs. Clinton, added, "If I have to become a fund-raiser to get an invite to her opening up the files, I'll do it."

When Mrs. Clinton started to talk openly about U.F.O.s and government disclosure in her 2016 campaign, some activists traced the remarks back to the 1995 meeting with Mr. Rockefeller.

To this subset of Americans who say the government is covering up what it knows about aliens, and who are incredibly vocal on social media, Mrs. Clinton's discussion of extraterrestrials signaled an important turn.

Other activists do not care as much about Mrs. Clinton's vow to "open the files," but do want prominent politicians to seriously acknowledge that humans may not be the only intelligent life in the universe. A major victory, some say, would be for the candidates to be asked about the topic in a presidential debate.

"It shouldn't be a source of embarrassment to discuss it," said Christopher Mellon, a former intelligence official at the Defense Department and the Senate Intelligence Committee. "We should be humble in terms of recognizing the extreme limits of our own understanding of physics and the universe."

2 Navy Airmen and an Object That 'Accelerated Like Nothing I've Ever Seen'

BY HELENE COOPER, LESLIE KEAN AND RALPH BLUMENTHAL | DEC. 16, 2017

The following recounts an incident in 2004 that advocates of research into U.F.O.s have said is the kind of event worthy of more investigation, and that was studied by a Pentagon program that investigated U.F.O.s. Experts caution that earthly explanations often exist for such incidents, and that not knowing the explanation does not mean that the event has interstellar origins.

CMDR. DAVID FRAVOR and Lt. Cmdr. Jim Slaight were on a routine training mission 100 miles out into the Pacific when the radio in each of their F/A-18F Super Hornets crackled: An operations officer aboard the U.S.S. Princeton, a Navy cruiser, wanted to know if they were carrying weapons.

"Two CATM-9s," Commander Fravor replied, referring to dummy missiles that could not be fired. He had not been expecting any hostile exchanges off the coast of San Diego that November afternoon in 2004.

Commander Fravor, in a recent interview with The New York Times, recalled what happened next. Some of it is captured in a video made public by officials with a Pentagon program that investigated U.F.O.s.

"Well, we've got a real-world vector for you," the radio operator said, according to Commander Fravor. For two weeks, the operator said, the Princeton had been tracking mysterious aircraft. The objects appeared suddenly at 80,000 feet, and then hurtled toward the sea, eventually stopping at 20,000 feet and hovering. Then they either dropped out of radar range or shot straight back up.

The radio operator instructed Commander Fravor and Commander Slaight, who has given a similar account, to investigate.

M. SCOTT BRAUER FOR THE NEW YORK TIMES

David Fravor, at his home in Windham, N.H., is a former Navy pilot who says he was "pretty weirded out" by an unexplained episode over the Pacific. His story has captured the attention of a Pentagon program investigating U.F.O.s.

The two fighter planes headed toward the objects. The Princeton alerted them as they closed in, but when they arrived at "merge plot" with the object — naval aviation parlance for being so close that the Princeton could not tell which were the objects and which were the fighter jets — neither Commander Fravor nor Commander Slaight could see anything at first. There was nothing on their radars, either.

Then, Commander Fravor looked down to the sea. It was calm that day, but the waves were breaking over something that was just below the surface. Whatever it was, it was big enough to cause the sea to churn.

Hovering 50 feet above the churn was an aircraft of some kind — whitish — that was around 40 feet long and oval in shape. The craft was jumping around erratically, staying over the wave disturbance but not moving in any specific direction, Commander Fravor said. The disturbance looked like frothy waves and foam, as if the water were boiling.

ALIENS AND THE U.S. GOVERNMENT

Commander Fravor began a circular descent to get a closer look, but as he got nearer the object began ascending toward him. It was almost as if it were coming to meet him halfway, he said.

Commander Fravor abandoned his slow circular descent and headed straight for the object.

But then the object peeled away. "It accelerated like nothing I've ever seen," he said in the interview. He was, he said, "pretty weirded out."

The two fighter jets then conferred with the operations officer on the Princeton and were told to head to a rendezvous point 60 miles away, called the cap point, in aviation parlance.

They were en route and closing in when the Princeton radioed again. Radar had again picked up the strange aircraft.

"Sir, you won't believe it," the radio operator said, "but that thing is at your cap point."

"We were at least 40 miles away, and in less than a minute this thing was already at our cap point," Commander Fravor, who has since retired from the Navy, said in the interview.

By the time the two fighter jets arrived at the rendezvous point, the object had disappeared.

The fighter jets returned to the Nimitz, where everyone on the ship had learned of Commander Fravor's encounter and was making fun of him.

Commander Fravor's superiors did not investigate further and he went on with his career, deploying to the Persian Gulf to provide air support to ground troops during the Iraq war. But he does remember what he said that evening to a fellow pilot who asked him what he thought he had seen.

"I have no idea what I saw," Commander Fravor replied to the pilot. "It had no plumes, wings or rotors and outran our F-18s."

But, he added, "I want to fly one."

Glowing Auras and 'Black Money': The Pentagon's Mysterious U.F.O. Program

BY HELENE COOPER, RALPH BLUMENTHAL AND LESLIE KEAN | DEC. 16, 2017

WASHINGTON — In the $600 billion annual Defense Department budgets, the $22 million spent on the Advanced Aerospace Threat Identification Program was almost impossible to find.

Which was how the Pentagon wanted it.

For years, the program investigated reports of unidentified flying objects, according to Defense Department officials, interviews with program participants and records obtained by The New York Times. It was run by a military intelligence official, Luis Elizondo, on the fifth floor of the Pentagon's C Ring, deep within the building's maze.

The Defense Department has never before acknowledged the existence of the program, which it says it shut down in 2012. But its backers say that, while the Pentagon ended funding for the effort at that time, the program remains in existence. For the past five years, they say, officials with the program have continued to investigate episodes brought to them by service members, while also carrying out their other Defense Department duties.

The shadowy program — parts of it remain classified — began in 2007, and initially it was largely funded at the request of Harry Reid, the Nevada Democrat who was the Senate majority leader at the time and who has long had an interest in space phenomena. Most of the money went to an aerospace research company run by a billionaire entrepreneur and longtime friend of Mr. Reid's, Robert Bigelow, who is currently working with NASA to produce expandable craft for humans to use in space.

On CBS's "60 Minutes" in May, Mr. Bigelow said he was "absolutely convinced" that aliens exist and that U.F.O.s have visited Earth.

AL DRAGO/THE NEW YORK TIMES

Harry Reid, the former Senate majority leader, has had a longtime interest in space phenomena.

Working with Mr. Bigelow's Las Vegas-based company, the program produced documents that describe sightings of aircraft that seemed to move at very high velocities with no visible signs of propulsion, or that hovered with no apparent means of lift.

Officials with the program have also studied videos of encounters between unknown objects and American military aircraft — including one released in August of a whitish oval object, about the size of a commercial plane, chased by two Navy F/A-18F fighter jets from the aircraft carrier Nimitz off the coast of San Diego in 2004.

Mr. Reid, who retired from Congress this year, said he was proud of the program. "I'm not embarrassed or ashamed or sorry I got this thing going," Mr. Reid said in a recent interview in Nevada. "I think it's one of the good things I did in my congressional service. I've done something that no one has done before."

Two other former senators and top members of a defense spending subcommittee — Ted Stevens, an Alaska Republican, and Daniel K. Inouye, a Hawaii Democrat — also supported the program. Mr. Stevens died in 2010, and Mr. Inouye in 2012.

While not addressing the merits of the program, Sara Seager, an astrophysicist at M.I.T., cautioned that not knowing the origin of an object does not mean that it is from another planet or galaxy. "When people claim to observe truly unusual phenomena, sometimes it's worth investigating seriously," she said. But, she added, "what people sometimes don't get about science is that we often have phenomena that remain unexplained."

James E. Oberg, a former NASA space shuttle engineer and the author of 10 books on spaceflight who often debunks U.F.O. sightings, was also doubtful. "There are plenty of prosaic events and human perceptual traits that can account for these stories," Mr. Oberg said. "Lots of people are active in the air and don't want others to know about it. They are happy to lurk unrecognized in the noise, or even to stir it up as camouflage."

Still, Mr. Oberg said he welcomed research. "There could well be a pearl there," he said.

In response to questions from The Times, Pentagon officials this month acknowledged the existence of the program, which began as part of the Defense Intelligence Agency. Officials insisted that the effort had ended after five years, in 2012.

"It was determined that there were other, higher priority issues that merited funding, and it was in the best interest of the DoD to make a change," a Pentagon spokesman, Thomas Crosson, said in an email, referring to the Department of Defense.

But Mr. Elizondo said the only thing that had ended was the effort's government funding, which dried up in 2012. From then on, Mr. Elizondo said in an interview, he worked with officials from the Navy and the C.I.A. He continued to work out of his Pentagon office until this past October, when he resigned to protest what he characterized as excessive secrecy and internal opposition.

JUSTIN T. GELLERSON FOR THE NEW YORK TIMES

Luis Elizondo, who led the Pentagon effort to investigate U.F.O.s until October. He resigned to protest what he characterized as excessive secrecy and internal opposition to the program.

"Why aren't we spending more time and effort on this issue?" Mr. Elizondo wrote in a resignation letter to Defense Secretary Jim Mattis.

Mr. Elizondo said that the effort continued and that he had a successor, whom he declined to name.

U.F.O.s have been repeatedly investigated over the decades in the United States, including by the American military. In 1947, the Air Force began a series of studies that investigated more than 12,000 claimed U.F.O. sightings before it was officially ended in 1969. The project, which included a study code-named Project Blue Book, started in 1952, concluded that most sightings involved stars, clouds, conventional aircraft or spy planes, although 701 remained unexplained.

Robert C. Seamans Jr., the secretary of the Air Force at the time, said in a memorandum announcing the end of Project Blue Book that it "no longer can be justified either on the ground of national security or in the interest of science."

Mr. Reid said his interest in U.F.O.s came from Mr. Bigelow. In 2007, Mr. Reid said in the interview, Mr. Bigelow told him that an official with the Defense Intelligence Agency had approached him wanting to visit Mr. Bigelow's ranch in Utah, where he conducted research.

Mr. Reid said he met with agency officials shortly after his meeting with Mr. Bigelow and learned that they wanted to start a research program on U.F.O.s. Mr. Reid then summoned Mr. Stevens and Mr. Inouye to a secure room in the Capitol.

"I had talked to John Glenn a number of years before," Mr. Reid said, referring to the astronaut and former senator from Ohio, who died in 2016. Mr. Glenn, Mr. Reid said, had told him he thought that the federal government should be looking seriously into U.F.O.s, and should be talking to military service members, particularly pilots, who had reported seeing aircraft they could not identify or explain.

The sightings were not often reported up the military's chain of command, Mr. Reid said, because service members were afraid they would be laughed at or stigmatized.

The meeting with Mr. Stevens and Mr. Inouye, Mr. Reid said, "was one of the easiest meetings I ever had."

He added, "Ted Stevens said, 'I've been waiting to do this since I was in the Air Force.'" (The Alaska senator had been a pilot in the Army's air force, flying transport missions over China during World War II.)

During the meeting, Mr. Reid said, Mr. Stevens recounted being tailed by a strange aircraft with no known origin, which he said had followed his plane for miles.

None of the three senators wanted a public debate on the Senate floor about the funding for the program, Mr. Reid said. "This was so-called black money," he said. "Stevens knows about it, Inouye knows about it. But that was it, and that's how we wanted it." Mr. Reid was referring to the Pentagon budget for classified programs.

Contracts obtained by The Times show a congressional appropriation of just under $22 million beginning in late 2008 through 2011.

The money was used for management of the program, research and assessments of the threat posed by the objects.

The funding went to Mr. Bigelow's company, Bigelow Aerospace, which hired subcontractors and solicited research for the program.

Under Mr. Bigelow's direction, the company modified buildings in Las Vegas for the storage of metal alloys and other materials that Mr. Elizondo and program contractors said had been recovered from unidentified aerial phenomena. Researchers also studied people who said they had experienced physical effects from encounters with the objects and examined them for any physiological changes. In addition, researchers spoke to military service members who had reported sightings of strange aircraft.

"We're sort of in the position of what would happen if you gave Leonardo da Vinci a garage-door opener," said Harold E. Puthoff, an engineer who has conducted research on extrasensory perception for the C.I.A. and later worked as a contractor for the program. "First of all, he'd try to figure out what is this plastic stuff. He wouldn't know anything about the electromagnetic signals involved or its function."

The program collected video and audio recordings of reported U.F.O. incidents, including footage from a Navy F/A-18 Super Hornet showing an aircraft surrounded by some kind of glowing aura traveling at high speed and rotating as it moves. The Navy pilots can be heard trying to understand what they are seeing. "There's a whole fleet of them," one exclaims. Defense officials declined to release the location and date of the incident.

"Internationally, we are the most backward country in the world on this issue," Mr. Bigelow said in an interview. "Our scientists are scared of being ostracized, and our media is scared of the stigma. China and Russia are much more open and work on this with huge organizations within their countries. Smaller countries like Belgium, France, England and South American countries like Chile are more open, too. They are proactive and willing to discuss this topic, rather than being held back by a juvenile taboo."

ISAAC BREKKEN FOR THE NEW YORK TIMES

Robert Bigelow, a billionaire entrepreneur and longtime friend of Mr. Reid, received most of the money allocated for the Pentagon program. On CBS's "60 Minutes" in May, Mr. Bigelow said he was "absolutely convinced" that aliens exist and that U.F.O.s have visited Earth.

By 2009, Mr. Reid decided that the program had made such extraordinary discoveries that he argued for heightened security to protect it. "Much progress has been made with the identification of several highly sensitive, unconventional aerospace-related findings," Mr. Reid said in a letter to William Lynn III, a deputy defense secretary at the time, requesting that it be designated a "restricted special access program" limited to a few listed officials.

A 2009 Pentagon briefing summary of the program prepared by its director at the time asserted that "what was considered science fiction is now science fact," and that the United States was incapable of defending itself against some of the technologies discovered. Mr. Reid's request for the special designation was denied.

Mr. Elizondo, in his resignation letter of Oct. 4, said there was a need for more serious attention to "the many accounts from the Navy and other services of unusual aerial systems interfering with military weapon platforms and displaying beyond-next-generation capabilities." He expressed his frustration with the limitations placed on the program, telling Mr. Mattis that "there remains a vital need to ascertain capability and intent of these phenomena for the benefit of the armed forces and the nation."

Mr. Elizondo has now joined Mr. Puthoff and another former Defense Department official, Christopher K. Mellon, who was a deputy assistant secretary of defense for intelligence, in a new commercial venture called To the Stars Academy of Arts and Science. They are speaking publicly about their efforts as their venture aims to raise money for research into U.F.O.s.

In the interview, Mr. Elizondo said he and his government colleagues had determined that the phenomena they had studied did not seem to originate from any country. "That fact is not something any government or institution should classify in order to keep secret from the people," he said.

For his part, Mr. Reid said he did not know where the objects had come from. "If anyone says they have the answers now, they're fooling themselves," he said. "We do not know."

But, he said, "we have to start someplace."

On the Trail of a Secret Pentagon U.F.O. Program

TIMES INSIDER | BY RALPH BLUMENTHAL | DEC. 18, 2017

Times Insider delivers behind-the-scenes insights into how news, features and opinion come together at The New York Times.

OUR READERS ARE plenty interested in unidentified flying objects. We know that from the huge response to our front-page Sunday article (published online just after noon on Saturday) revealing a secret Pentagon program to investigate U.F.O.s. The piece, by the Pentagon correspondent Helene Cooper, the author Leslie Kean and myself — a contributor to The Times after a 45-year staff career — has dominated the most emailed and most viewed lists since.

So how does a story on U.F.O.s get into The New York Times? Not easily, and only after a great deal of vetting, I assure you.

The journey began two and a half months ago with a tip to Leslie, who has long reported on U.F.O.s and published a 2010 New York Times best seller, "UFOs: Generals, Pilots and Government Officials Go on the Record." At a confidential meeting Oct. 4 in a Pentagon City hotel with several present and former intelligence officials and a defense contractor, she met Luis Elizondo, the director of a Pentagon program she had never heard of: the Advanced Aerospace Threat Identification Program.

She learned it was a secret effort, funded at the initiative of the then Senate majority leader, Harry Reid, starting in 2007, to investigate aerial threats including what the military preferred to call "unidentified aerial phenomena" or just "objects." This was big news because the United States military had announced as far back as 1969 that U.F.O.s were not worth studying. Leslie also learned that Mr. Elizondo had just resigned to protest what he characterized as excessive secrecy and internal opposition — the reason for the meeting.

She spent hours with him reviewing unclassified documents, for the $22 million program operated largely "in the white" (that is, not under special restricted access), but hidden in the huge defense budget, with only parts of it classified. A few days later Mr. Elizondo and others there — including Harold E. Puthoff, an engineer who has conducted research on extrasensory perception for the C.I.A. and later worked as a contractor on the program, and Christopher K. Mellon, a former deputy assistant secretary of defense for intelligence — announced they were joining a new commercial venture, To the Stars Academy of Arts and Science, to raise money for research into U.F.O.s. Leslie wrote it up for the Huffington Post with scant details of the program.

I had known Leslie for years, and she told me this looked like a story for The Times. I agreed. Leslie and I met with Mr. Elizondo in Philadelphia on Oct. 31. Three days later, I emailed the executive editor, Dean Baquet, about "a sensational and highly confidential time-sensitive story" that I said "involves a senior U.S. intelligence official who abruptly quit last month" exposing "a deeply secret program, long mythologized but now confirmed."

He alerted Mark Mazzetti, the investigations editor in the Washington bureau. Leslie and I briefed him in New York on Nov. 7. We assured him there were no anonymous sources; everyone was on the record. After discussions in Washington and New York, Helene joined our team. The Washington bureau chief, Elisabeth Bumiller, would be our editor. On Nov. 17, we three met Mr. Elizondo in a nondescript Washington hotel where he sat with his back to the wall, keeping an eye on the door.

Carl Hulse, The Times's chief Washington correspondent, was well acquainted with Mr. Reid and helped arrange an interview for Helene. She flew to Las Vegas on Dec. 5 and met with the former senator, who confirmed the program with details, saying, "I'm not embarrassed or ashamed or sorry I got this going."

Leslie interviewed the aerospace magnate Robert Bigelow, who also confirmed his participation, saying Americans were being held back from serious research into U.F.O.'s by "a juvenile taboo." And I interviewed a prominent skeptic for perspective.

It was important that we not take anything on faith. This field attracts zealots as well as debunkers, and many Americans remain deeply skeptical that the phenomenon exists as popularly portrayed. In draft after draft, we took pains to let the investigation speak for itself, without bias.

Helene met with a Pentagon spokesperson on Dec. 8 for a response to the information we had gathered. The answer came swiftly. There had been a program to investigate U.F.O.s, but it ended in 2012 after five years, the Defense Department insisted.

Our reporting suggested it continues, largely unfunded, to the present. And that's what we wrote.

CHAPTER 3

The Search for Extraterrestrial Life

As technology has progressed, so too has our ability to hunt for alien life. Scientists and astronomers have celebrated recent developments that have shown evidence of water on Mars and inhabitable, Earth-like planets in solar systems not too far removed from our own. The presence of Oumuamua, a large object that was detected in our solar system in 2017, was thought to be possible evidence of other intelligent life. The articles in this section speak to the science and the continued hunt for life on other planets.

The Intelligent-Life Lottery

COLUMN | BY GEORGE JOHNSON | AUG. 18, 2014

ALMOST 20 YEARS AGO, in the pages of an obscure publication called Bioastronomy News, two giants in the world of science argued over whether SETI — the Search for Extraterrestrial Intelligence — had a chance of succeeding. Carl Sagan, as eloquent as ever, gave his standard answer. With billions of stars in our galaxy, there must be other civilizations capable of transmitting electromagnetic waves. By scouring the sky with radio telescopes, we just might intercept a signal.

But Sagan's opponent, the great evolutionary biologist Ernst Mayr, thought the chances were close to zero. Against Sagan's stellar billions, he posed his own astronomical numbers: Of the billions of

CARL WIENS

species that have lived and died since life began, only one — Homo sapiens — had developed a science, a technology, and the curiosity to explore the stars. And that took about 3.5 billion years of evolution. High intelligence, Mayr concluded, must be extremely rare, here or anywhere. Earth's most abundant life form is unicellular slime.

Since the debate with Sagan, more than 1,700 planets have been discovered beyond the solar system — 700 just this year. Astronomers recently estimated that one of every five sunlike stars in the

Milky Way might be orbited by a world capable of supporting some kind of life.

That is about 40 billion potential habitats. But Mayr, who died in 2005 at the age of 100, probably wouldn't have been impressed. By his reckoning, the odds would still be very low for anything much beyond slime worlds. No evidence has yet emerged to prove him wrong.

Maybe we're just not looking hard enough. Since SETI began in the early 1960s, it has struggled for the money it takes to monitor even a fraction of the sky. In an online essay for The Conversation last week, Seth Shostak, the senior astronomer at the SETI Institute, lamented how little has been allocated for the quest — just a fraction of NASA's budget.

"If you don't ante up," he wrote, "you will never win the jackpot. And that is a question of will."

Three years ago, SETI's Allen Telescope Array in Northern California ran out of money and was closed for a while. Earlier this month, it was threatened by wildfire — another reminder of the precariousness of the search.

It has been more than 3.5 billion years since the first simple cells arose, and it took another billion years or so for some of them to evolve and join symbiotically into primitive multicellular organisms. These biochemical hives, through random mutations and the blind explorations of evolution, eventually led to creatures with the ability to remember, to anticipate and — at least in the case of humans — to wonder what it is all about.

Every step was a matter of happenstance, like the arbitrary combination of numbers — 3, 12, 31, 34, 51 and 24 — that qualified a Powerball winner for a $90 million prize this month. Some unknowing soul happened to enter a convenience store in Rifle, Colo., and — maybe with change from buying gasoline or a microwaved burrito — purchase a ticket just as the machine was about to spit out those particular numbers.

According to the Powerball website, the chance of winning the grand prize is about one in 175 million. The emergence of humanlike intelligence, as Mayr saw it, was about as likely as if a Powerball winner kept buying tickets and — round after round — hit a bigger jackpot each time. One unlikelihood is piled on another, yielding a vanishingly rare event.

In one of my favorite books, "Wonderful Life," Stephen Jay Gould celebrated what he saw as the unlikelihood of our existence. Going further than Mayr, he ventured that if a slithering creature called Pikaia gracilens had not survived the Cambrian extinction, about half a billion years ago, the entire phylum called Chordata, which includes us vertebrates, might never have existed.

Gould took his title from the Frank Capra movie in which George Bailey gets to see what the world might have been like without him — idyllic Bedford Falls is replaced by a bleak, Dickensian Pottersville.

For Gould, the fact that any of our ancestral species might easily have been nipped in the bud should fill us "with a new kind of amazement" and "a frisson for the improbability of the event" — a fellow agnostic's version of an epiphany.

"We came *this close* (put your thumb about a millimeter away from your index finger), thousands and thousands of times, to erasure by the veering of history down another sensible channel," he wrote. "Replay the tape a million times," he proposed, "and I doubt that anything like Homo sapiens would ever evolve again. It is, indeed, a wonderful life."

Other biologists have disputed Gould's conclusion. In the course of evolution, eyes and multicellularity arose independently a number of times. So why not vertebrae, spinal cords and brains? The more bags of tricks an organism has at its disposal, the greater its survival power may be. A biological arms race ensues, with complexity ratcheted ever higher.

But those occasions are rare. Most organisms, as Daniel Dennett put it in "Darwin's Dangerous Idea," seem to have "hit upon a relatively simple solution to life's problems at the outset and, having nailed it a billion years ago, have had nothing much to do in the way of design work ever since." Our appreciation of complexity, he wrote, "may well be just an aesthetic preference."

In "Five Billion Years of Solitude," by Lee Billings, published last year, the author visited Frank Drake, one of the SETI pioneers.

"Right now, there could well be messages from the stars flying right through this room," Dr. Drake told him. "Through you and me. And if we had the right receiver set up properly, we could detect them. I still get chills thinking about it."

He knew the odds of tuning in — at just the right frequency at the right place and time — were slim. But that just meant we needed to expand the search.

"We've been playing the lottery only using a few tickets," he said.

GEORGE JOHNSON writes the Raw Data column for The New York Times.

Should We Keep a Low Profile in Space?

OPINION | BY SETH SHOSTAK | MARCH 27, 2015

MOUNTAIN VIEW, CALIF. — For more than a half-century, a small group of astronomers has sought intelligent company among the stars. They've done so by turning large radio antennas skyward, hoping to eavesdrop on signals from an advanced society. It's a program known as SETI, the Search for Extraterrestrial Intelligence.

But now some researchers propose that we should do more than simply don headphones and await E.T.'s call: We should make serious efforts to encourage a response from putative aliens by deliberately transmitting our own messages. It's a simple idea, akin to tossing a bottle into the cosmic ocean. But recent arguments for what's termed active SETI have loosed a storm of controversy, one that has even washed into the halls of academe.

Why is this? Why has the sending of dispatches to worlds many trillions of miles distant suddenly become a hot-button issue? The simple answer is that there's now a perception that advertising our existence could be a mortal threat to the planet.

The reasoning is this: While no one has yet offered decisive proof for life beyond Earth, in the past two years astronomers have learned that tens of billions of habitable planets suffuse our galaxy. Consequently, to believe that only Earth has spawned intelligence is to insist that our world is the site of a miracle. That point of view rarely appeals to scientists.

The aliens could very well be out there. And that realization has spurred a call by some for broadcasts intended to elicit a communication from at least the nearest other star systems. But we know nothing of the aliens' possible motives or behavior. Therefore, it's conceivable that betraying our existence might prompt aggressive action from space.

Broadcasting is likened to "shouting in the jungle" — not a good idea when you don't know what's out there. The British physicist Stephen Hawking alluded to this danger by noting that on Earth, when less advanced societies drew the attention of those more advanced, the consequences for the former were seldom agreeable.

It's a worry we never used to have. Victorian-era scientists toyed with plans to use lanterns and burning pools of oil to contact postulated Martians. In the 1970s, NASA bolted greeting cards onto spacecraft that will leave our solar system and wander the vast reaches between the stars. The Pioneer and Voyager probes carry plaques and records with information about what humans look like and where Earth is, as well as a small sampling of our culture.

Those messages move at the speed of rockets. But in 1974, a three-minute encoded pictogram was transmitted using the large radio antenna at Arecibo, Puerto Rico. It moves at the speed of light, 20,000 times faster. More recent radio transmissions include a Beatles song beamed by NASA to the North Star, a Doritos advertisement launched to a planetary system in the Big Dipper, and a series of broadcasts sent to nearby stars using an antenna in Crimea.

When most people believed that aliens were no more than easy black hats for Hollywood, the idiosyncratic nature of these messages could be easily dismissed. But if cosmic company is a legitimate possibility, shouldn't we offer up something more edifying than pop music and snack food? A deliberate transmission should represent all of humanity — not short-circuit the important question of who will speak for Earth.

Consequently, recent conferences on the merits of active SETI have sought the advice of social scientists. Among their worries is whether to be up front about humanity's seamy side: Should we tell the extraterrestrials about war and injustice?

Personally, I think this concern is overwrought. Any society that can pick up our radio messages will be at a level of development at least centuries beyond our own. They would be no more incensed by our bad behavior than historians who learned that Babylonians attacked one another with spears. It seems naïve to imagine that, by shielding aliens from the less flattering aspects of humanity, we would somehow lessen any incentive to do us harm. If there's a danger, mincing words is unlikely to eliminate it.

A better approach is to note that the nearest intelligent extraterrestrials are likely to be at least dozens of light-years away. Even assuming that active SETI provokes a reply, it won't be breezy conversation. Simple back-and-forth exchanges would take decades. This suggests that we should abandon the "greeting card" format of previous signaling schemes, and offer the aliens Big Data.

For example, we could transmit the contents of the Internet. Such a large corpus — with its text, pictures, videos and sounds — would allow clever extraterrestrials to decipher much about our society, and even formulate questions that could be answered with the material in hand. Sending the web on its way would take months if a radio transmitter were used. A powerful laser, conveying bits much like an optical fiber, could launch these data in a few days.

Sending messages — even big ones — is technically feasible. However, there's still the highly controversial matter of whether to broadcast at all. Who decides? One could simply let the public weigh in, but doing so wouldn't address the security issue. Even if a majority is comfortable with a transmission, how does that mitigate the possible danger?

The inability to gauge this peril prompts some critics to argue that, given the possibly existential threat posed by active SETI, we should choose the side of caution. We should simply forbid powerful transmissions to the skies. Indeed, a small consortium of academics in California has drafted a petition urging this.

It's a wary approach. It's also poor insurance. Any extraterrestrials with technology advanced enough to threaten us will surely have antennas larger than our own, instruments that can pick up the television and radio signals broadcast willy-nilly since World War II. We are already shouting into the jungle, albeit with less volume than a deliberate signal. But the dangerous creatures may have good hearing.

Additionally, if we forbid high-powered transmitters aimed at the sky, we shut out such obvious future technologies as better radars for aviation and tracking dangerous asteroids. Do we really want to hamstring our descendants this way?

A decision to engage in active SETI has not been made. The benefit — learning our place in the cosmos — is only hypothetical, and so is the danger. But I, for one, would hesitate to let a paranoia based on nothing more than conjecture shackle the activities of our children and our children's children. The universe beckons, and we can do better than to declare that future generations should endlessly tremble at the sight of the stars.

SETH SHOSTAK is the director of the Center for SETI Research at the SETI Institute, and a host of the radio program "Big Picture Science."

The Flip Side of Optimism About Life on Other Planets

COLUMN | BY DENNIS OVERBYE | AUG. 3, 2015

IF YOU DREAM of close encounters of the alien kind, this has been a hopeful summer.

In July, on the 46th anniversary of the first moon landing, Yuri Milner, the Russian Internet entrepreneur and philanthropist, said he would spend $100 million over the next decade on the search for alien signals, known as SETI, giving the field a financial stability and access to telescopes it had never had. That same week, NASA announced the discovery of what might be the most Earthlike planet yet beyond the solar system, Kepler 452b, a mere 1,400 light-years from here.

In a news conference accompanying Mr. Milner's announcement, Geoffrey Marcy, a planet hunter from the University of California, Berkeley, noted that "the universe is apparently bulging at the seams with ingredients for biology."

He said he would bet Yuri Milner's house, reportedly also worth $100 million, that there is at least microbial life out there.

You might think the discovery of microbes on Mars or fish in the oceans of Jupiter's moon Europa would have scientists dancing in the streets. And you would probably be right.

But not everyone agrees that it would be such good news. For at least one prominent thinker, it would be a "crushing blow."

That would be Nick Bostrom, a philosopher at the University of Oxford and director of the Future of Humanity Institute there, one of the great pessimists of this or any other age.

In an article published in Technology Review in 2008, Professor Bostrom declared that it would be a really bad sign for the future of humanity if we found even a microbe clinging to a rock on Mars. "Dead rocks and lifeless sands would lift my spirit," he wrote.

Why?

ELWOOD SMITH

It goes back to a lunch in 1950 in Los Alamos, N.M., the birthplace of the atomic bomb. The subject was flying saucers and interstellar travel. The physicist Enrico Fermi blurted out a question that has become famous among astronomers: "Where is everybody?"

The fact that there was no evidence outside supermarket tabloids that aliens had ever visited Earth convinced Fermi that interstellar travel was impossible. It would simply take too long to get anywhere.

The argument was expanded by scientists like Michael Hart and Frank Tipler, who concluded that extraterrestrial technological civilizations simply didn't exist.

The logic is simple. Imagine that one million years from now Earthlings launch a robot to Alpha Centauri, the closest star system to our own. It gets there in a few years, and a million years later sends off probes to two other star systems. A million years after that, each of those sends off two more probes. Even allowing for generous travel times, in 100 million years roughly a nonillion stars (10^{30}) could be visited. The galaxy contains maybe 200 billion stars, so each could be visited more than a trillion times in this robot crisscrossing.

The interstellar probe part of this is not so crazy, by the way. Serious people are already contemplating sending a probe to another star, using technology that could be achievable in the near future. See, for example, the Defense Advanced Research Projects Agency and its 100-Year Starship Study.

There are billions of potentially habitable planets in the galaxy, moreover. If only a small fraction of these develop life and technology, that would be enough to turn the whole galaxy into Times Square.

The Milky Way is 10 billion years old. So where are those aliens or their artifacts? We've found zilch. If life is so easy, someone from somewhere must have come calling by now. This is known as the Fermi paradox.

There are many loopholes in this argument, including the possibility that we wouldn't sufficiently recognize alien life if it camped in our front yards. The simplest explanation, Dr. Bostrom and others say, is that there are no other spacefaring civilizations.

There must be something, he concludes, that either stops life from starting at all, or shuts it down before it can conquer the stars. He calls it the Great Filter.

You can imagine all kinds of bottlenecks in the evolution of life and civilization — from the need for atoms to first combine into strands of RNA, the genetic molecule that plays Robin to DNA's Batman, to nuclear war, climate change or a mishap of genetic engineering — that could constitute a Great Filter.

The big question for Professor Bostrom is whether the Great Filter is in our past or our future, and for the answer he looks to the stars. If there is nothing else out there, then maybe we have survived whatever it is. As bizarre as it sounds, we are the first ones in the neighborhood to have run the cosmic obstacle course.

If there is company out there, it means the Great Filter is ahead of us. We are doomed.

This is a staggeringly existential piece of knowledge to have obtained at what seems to be a tender age as a species, based on a cursory examination of a sliver of our cosmic neighborhood. It is also a truly brave exercise of the power of human reason.

Maybe too brave. But there is a precedent of sorts in an old riddle known as Olbers' Paradox, after Heinrich Wilhelm Olbers, a 19th-century amateur astronomer who enunciated a problem that had bothered some astronomers since the 16th century: Why is the sky dark at night? In an infinite eternal universe, every line of sight would end on a star, the thinking went, and even dust clouds would glow as bright as day.

Luminaries as disparate as the Scottish physicist Lord Kelvin and the writer Edgar Allan Poe suggested that the dark night sky was a clue to the fact that the universe is finite, at least in time, and had a beginning, a notion now cemented by the Big Bang.

If Olbers saw the dawn of time, perhaps Fermi and Bostrom have seen the sunset. We should hardly be surprised. Nothing lasts forever. The fathers of SETI, Carl Sagan and Frank Drake, stressed that a key unknown element in their equations was the average lifetime of technological civilizations. Too short a lifetime would eliminate the possibility of overlapping civilizations. Forget about the mythical brotherhood of the galaxy. The Klingons left the building long ago.

The best we could have hoped for was to be another evolutionary phase in the zigzag development of earthly life on the way to who knows what. But in a few billion years, the sun will die, and so will the earth, and our descendants — if they are still on it. The universe will not remember us or Shakespeare or Homer.

We can't really blame Professor Bostrom for that. But he has a history of disturbing thoughts. In 2003, he argued that we were probably all living in a computer simulation, something he said would be easy for "technologically mature" civilizations to do.

What his and other sci-fi-style calculations have in common is that they are extrapolations, of the doubling of chip capacity decreed by

Moore's Law in the case of computer simulations, or the doubling of space probes over the eons. Believe them at your peril. Chips can't get smaller forever. Untended machines far, far from home break or forget why they are there. Apple can't keep doubling iPhone sales eternally.

As the great science writer and biologist Lewis Thomas liked to say, we are an ignorant species. This is why we do the experiment.

DENNIS OVERBYE writes the Out There column for The New York Times.

Yes, There Have Been Aliens

OPINION | BY ADAM FRANK | JUNE 10, 2016

LAST MONTH ASTRONOMERS from the Kepler spacecraft team announced the discovery of 1,284 new planets, all orbiting stars outside our solar system. The total number of such "exoplanets" confirmed via Kepler and other methods now stands at more than 3,000.

This represents a revolution in planetary knowledge. A decade or so ago the discovery of even a single new exoplanet was big news. Not anymore. Improvements in astronomical observation technology have moved us from retail to wholesale planet discovery. We now know, for example, that every star in the sky likely hosts at least one planet.

But planets are only the beginning of the story. What everyone wants to know is whether any of these worlds has aliens living on it. Does our newfound knowledge of planets bring us any closer to answering that question?

A little bit, actually, yes. In a paper published in the May issue of the journal Astrobiology, the astronomer Woodruff Sullivan and I show that while we do not know if any advanced extraterrestrial civilizations currently exist in our galaxy, we now have enough information to conclude that they almost certainly existed at some point in cosmic history.

Among scientists, the probability of the existence of an alien society with which we might make contact is discussed in terms of something called the Drake equation. In 1961, the National Academy of Sciences asked the astronomer Frank Drake to host a scientific meeting on the possibilities of "interstellar communication." Since the odds of contact with alien life depended on how many advanced extraterrestrial civilizations existed in the galaxy, Drake identified seven factors on which that number would depend, and incorporated them into an equation.

The first factor was the number of stars born each year. The second was the fraction of stars that had planets. After that came the number of planets per star that traveled in orbits in the right locations for

GÉRARD DUBOIS

life to form (assuming life requires liquid water). The next factor was the fraction of such planets where life actually got started. Then came factors for the fraction of life-bearing planets on which intelligence and advanced civilizations (meaning radio signal-emitting) evolved. The final factor was the average lifetime of a technological civilization.

Drake's equation was not like Einstein's $E=mc^2$. It was not a statement of a universal law. It was a mechanism for fostering organized discussion, a way of understanding what we needed to know to answer the question about alien civilizations. In 1961, only the first factor — the number of stars born each year — was understood. And that level of ignorance remained until very recently.

That's why discussions of extraterrestrial civilizations, no matter how learned, have historically boiled down to mere expressions of hope or pessimism. What, for example, is the fraction of planets that form life? Optimists might marshal sophisticated molecular biological models to argue for a large fraction. Pessimists then cite their own scientific data to argue for a fraction closer to 0. But with only one example of a life-bearing planet (ours), it's hard to know who is right.

Or consider the average lifetime of a civilization. Humans have been using radio technology for only about 100 years. How much longer will our civilization last? A thousand more years? A hundred thousand more? Ten million more? If the average lifetime for a civilization is short, the galaxy is likely to be unpopulated most of the time. Once again, however, with only one example to draw from, it's back to a battle between pessimists and optimists.

But our new planetary knowledge has removed some of the uncertainty from this debate. Three of the seven terms in Drake's equation are now known. We know the number of stars born each year. We know that the percentage of stars hosting planets is about 100. And we also know that about 20 to 25 percent of those planets are in the right place for life to form. This puts us in a position, for the first time, to say something definitive about extraterrestrial civilizations — if we ask the right question.

In our recent paper, Professor Sullivan and I did this by shifting the focus of Drake's equation. Instead of asking how many civilizations currently exist, we asked what the probability is that ours is the only technological civilization that has ever appeared. By asking this question, we could bypass the factor about the average lifetime of a civilization. This left us with only three unknown factors, which we combined into one "biotechnical" probability: the likelihood of the creation of life, intelligent life and technological capacity.

You might assume this probability is low, and thus the chances remain small that another technological civilization arose. But what our calculation revealed is that even if this probability is assumed to be extremely low, the odds that we are not the first technological civilization are actually high. Specifically, unless the probability for evolving a civilization on a habitable-zone planet is less than one in 10 *billion trillion*, then we are not the first.

To give some context for that figure: In previous discussions of the Drake equation, a probability for civilizations to form of one in 10 billion per planet was considered highly pessimistic. According to our finding, even if you grant that level of pessimism, a *trillion* civilizations still would have appeared over the course of cosmic history.

In other words, given what we now know about the number and orbital positions of the galaxy's planets, the degree of pessimism required to doubt the existence, at some point in time, of an advanced extraterrestrial civilization borders on the irrational.

In science an important step forward can be finding a question that can be answered with the data at hand. Our paper did just this. As for the big question — whether any other civilizations currently exist — we may have to wait a long while for relevant data. But we should not underestimate how far we have come in a short time.

ADAM FRANK is an astrophysics professor at the University of Rochester, a co-founder of NPR's 13.7 Cosmos and Culture blog and the author of "About Time: Cosmology and Culture at the Twilight of the Big Bang."

Twinkle, Twinkle Little Trappist

EDITORIAL | BY THE NEW YORK TIMES | FEB. 24, 2017

SO, WE MAY have been looking for alien life in the wrong place! Not long ago, scientists scouring the cosmos for Earth-like planets with the right stuff to generate life were looking around sun-like stars. It turns out that the first such planets they've found — seven of them — are circling something quite different: what scientists call an "ultracool dwarf" in their ultracool terminology, though in this case the reference is to the temperature of a dim star barely one-twelfth the mass of the sun.

The discovery is enormously exciting, for several reasons. One is that the little star, which in their whimsical way the scientists named Trappist-1 after the telescope in Chile initially used to study it, is a mere 40 light-years from Earth, which is next door in cosmic terms. The search for alien life can now start far sooner than anticipated, especially with new telescopes about to come into service, and some answers might be available within a decade.

Then there's the fact that cool red dwarfs like Trappist-1 are the most common type of star, so there are probably many more potentially life-supporting worlds out there than were previously suspected. Astronomers have always presumed that other stars must have their planets, but it was only in 1995 that an exoplanet — one orbiting a star other than the sun — was confirmed. More than 3,400 of these have been discovered since.

The Trappist-1 cluster, however, is the first discovery of planets that are about the size of Earth and might have the right composition and temperature to have oceans of liquid water, and therefore, possibly, life. The planets were discovered by measuring dips in the light emitted by Trappist-1, which enabled astronomers to calculate their number and size. The next step will be to observe the planets for signs of the gases that would indicate life exists on them. In the meantime,

NASA

Seven Earth-size planets orbit a dwarf star named Trappist-1 about 40 light-years from Earth.

astronomers will be checking other ultracool dwarfs to see what's orbiting around them.

What makes this story so irresistible is the mystery and allure of the cosmos that all of us know from the first time we looked up at the stars. The article in the journal Nature announcing the discovery, signed by a large team of astronomers led by Michaël Gillon of the University of Liège in Belgium, began with the simple declaration that searching for Earth-like exoplanets is "one aim of modern astronomy." There was no effort, and no need, to further justify the enormous commitment of resources, ingenuity, time and effort in a project that, on the face of it, has no obvious commercial or practical benefit.

There is always the possibility of collateral benefits, of course, but none could be greater than finding out whether anyone else is out there.

A New Exoplanet May Be Most Promising Yet in Search for Life

BY DENNIS OVERBYE | APRIL 19, 2017

A PRIME PLANET LISTING has just appeared on the cosmic real estate market, possibly the most promising place yet to search for signs of life beyond the solar system, the astronomers who discovered it say.

It is a rocky orb about one and a half times the size of Earth, about 40 light years from here. It circles a dwarf star known as LHS 1140 every 25 days, an orbit that puts it in the "Goldilocks" zone where temperatures are conducive to liquid water and perhaps life as we know it.

It is close enough that astronomers are hopeful that with the next generation of big telescopes, they will be able to probe its atmosphere for signs of water or other evidence of suitability for life.

"This planet is really close to us: If we shrank the Milky Way to the size of the United States, LHS 1140 and the sun would fit inside Central Park," David Charbonneau, of the Harvard-Smithsonian Center for Astrophysics, said in an email.

His colleague Jason Dittmann, who led the discovery team and is lead author of a paper published on Wednesday in Nature, said in a statement, "This is the most exciting exoplanet I've seen in the last decade."

The planet was discovered by the MEarth-South survey at the Cerro Tololo Inter-American Observatory in Chile, an array of small telescopes that looks for the dips in starlight when planets pass in front of nearby stars.

The depth of the dip told them how big the new planet is. Then they determined that it was about six times as massive as Earth by using a spectrograph called Harps, for High Accuracy Radial velocity Planet Searcher, at the European Southern Observatory, also in Chile, to measure how much the planet perturbed its home star. The resulting density puts the little world into a rapidly growing class called "superEarths."

The star LHS 1140 is about one-fifth the size of our sun. In its close orbit, the planet receives about half as much energy as Earth does from its own sun, enough for a microbe or something more complicated to make a living.

This discovery continues a recent run of promising new planets circling nearby dwarf stars. Last summer there was the discovery of Proxima b, the nearest star to us, only 4.2 light years from here.

In February astronomers discovered a system of seven Earth-size planets circling a dwarf star known as Trappist-1.

According to Dr. Charbonneau, who originated the MEarth system, red dwarf stars outnumber stars like our sun by about 10 to 1 in the 30-light-year bubble that constitutes our "block" in the cosmos.

About one in four of them have rocky planets in their habitable zones, according to work by Dr. Charbonneau's former student Courtney Dressing, now at the California Institute of Technology.

Once upon a time, such planets were not looked upon favorably in the extraterrestrial life sweepstakes, because they were almost undoubtedly tidally locked, keeping one side faced to its star and the other facing out in space. That would result in a burning hell on one side and eternal frostbite on the other, neither side suitable for life.

But recently astronomers have determined that if these planets have thick enough atmospheres, winds can distribute the heat around both hemispheres and make them livable.

"Now we love them," Sara Seager, a planetary expert at the Massachusetts Institute of Technology, said at a recent meeting on the origins of life sponsored by Harvard at the American Academy of Arts and Sciences in Cambridge. "If they have an atmosphere, they can harbor life."

Astronomers said that the new planet offered the best hope so far to test that proposition. When the planet crosses in front of LHS 1140 the atmosphere acts like a filter, leaving an imprint on the star's light that could betray the presence of water and other molecules important for life.

This will be a job for powerful new telescopes like the James Webb Space Telescope, due to be launched next year, or giant ground-based telescopes like the Giant Magellan and European Extremely Large telescopes, now being built in Chile, Dr. Charbonneau said.

Whether such planets actually have atmospheres is still controversial, however. When red dwarf stars are young, Dr. Charbonneau pointed out, they are ferociously luminous and might have blown away the planets' atmospheres or caused a runaway greenhouse, leaving them barren. But the LHS 1140 planet is heavy enough, he said, that it might have been able to retain its atmosphere or regenerate it by volcanic activity later on.

"But the key point is yes, these are really exciting ideas to test," he continued. "Do temperate, rocky M-dwarfs planets retain their atmospheres, and do they have life? This world enables those studies."

'Aliens' Asks: If the Universe Is So Vast, Where Is Everybody?

REVIEW | BY JENNIFER SENIOR | MAY 24, 2017

ALIENS
The World's Leading Scientists on the Search for Extraterrestrial Life
Edited and With an Introduction by Jim Al-Khalili
232 pages. Picador. $25.

PERHAPS MY FAVORITE ESSAY in "Aliens: The World's Leading Scientists on the Search for Extraterrestrial Life" is by the astrobiologist Lewis Dartnell, who patiently explains why aliens would not come here to have sex with us or eat us for supper.

I can only assume that he gets these questions a lot.

Here are the answers, should you find these possibilities concerning: The likelihood that we'd be genetically compatible with aliens is terribly remote, which means that they'd almost certainly be immune to our sexual charms. For similar reasons, having to do with biochemistry, we'd be lousy refreshments for them — they would almost certainly lack the proper enzymes to digest us.

As a bonus, Dartnell goes on to reassure us why aliens wouldn't be especially interested in raiding our planet for raw materials, either (asteroids are a far easier source to mine); and if it were water they were after, they'd be far better off going to Europa, one of Jupiter's largest moons, which contains more water beneath its icy shell than all the oceans on Earth combined.

If you're interested in non-Earthly life, don't look to the movies, is his point.

You could argue that that's the general point of this modest, eccentric collection. Jim Al-Khalili, a quantum physicist and the editor of "Aliens," opens with a question asked by Enrico Fermi in 1950: If the universe is so vast, and its age so old, and its stars so plentiful, where is everybody?

I'm no marketing expert, but "Where Is Everybody?" strikes me as a far catchier title for this book than the one it has, and it's definitely more accurate. There really is nobody — so far — to write about. (Fighting words, I know. My hands hovered, spaceshiplike, for several minutes over the keyboard before committing that sentence to print.) This doesn't mean that life elsewhere doesn't exist. But it probably corresponds very little to what most of us have in mind, and not at all to the ooze-covered beasts of Ridley Scott's electric dreams.

One of the most consistent takeaways from this anthology is just how banal extraterrestrial life might be. Often, when entertaining the possibility of aliens, what we're really entertaining is the possibility of hardy microbes that can withstand extreme conditions, whether they're thermophiles (heat lovers), psychrophiles (cold lovers) or halophiles (salt lovers). Read enough of "Aliens," and you realize that the search for life is just as much about the most mundane aspects of biology as about the trippier questions of cosmology.

But it is also about philosophy. In this search, it helps to know what life *is*. Yet there's no consensus about how to answer this question, strangely. (At the risk of being too Clintonian, it depends on what your definition of "is" is.)

Nor do we know how life began. At some point, the Earth made the transition from chemistry to biology, yes, but we cannot "agree on a definition that separates the nonliving chemistry from life," as the geneticist Johnjoe McFadden puts it. (He then paraphrases the astronomer Fred Hoyle, who famously said that the odds of assembling something like a bacterium out of the primordial ooze were akin to the odds of a tornado's assembling a jumbo jet out of a junkyard heap as it sweeps through.)

There are scientists who will go so far as to say that life is a spectacular fluke. Not everyone, mind you: Researchers now estimate that there are one billion Earthlike exoplanets in the Milky Way. "To my mathematical brain, the numbers alone make thinking about aliens perfectly rational," Stephen Hawking has said.

But a powerful essay by the evolutionary biologist Matthew Cobb will make you wonder whether any form of multicellular life is far less likely than one in a billion. He points out that "there are more single-celled organisms alive on Earth than there are Earthlike planets in the observable universe"; that the number of single-celled organisms that have lived on this planet over the course of 3.8 billion years is beyond calculation; that these organisms have interacted "gazillions" of times (I love it when words of the appropriate magnitude desert even the experts). Yet we've never had a second instance of eukaryogenesis — that remarkable moment when one unicellular life form lodged inside another, forming something much more complex — in all this time.

Of course, there are researchers who dispute this theory and Cobb's reasoning. But you get the idea.

The experience of reading almost any anthology is a bit like traveling across the country in a rental car with only an FM radio for company. Sometimes you get Sinatra; other times you get Nickelback.

This collection has its share of Nickelback. One of its most disappointing essays is about aliens in science fiction, which manages, against stupefying odds, to contain just one interesting insight: that authors tend to be more concerned with physics than with biology. (How did those gigantic sandworms evolve on the desert planet in "Dune"?)

But the best of these essays are far out in more ways than one. The very first, by the cosmologist Martin Rees, notes that our best hope for interstellar travel isn't as humans, who don't live very long and require far too much fuel to get very far, but as post-humans, who will have made the Kurzweilian transition from organic to inorganic, from decaying mortals to silicon-based, eminently portable machines. He adds that alien intelligence, if we ever detect it, will also be in this form.

The final essay, by Seth Shostak, a senior astronomer at the SETI institute (short for Search for Extraterrestrial Intelligence), goes even further, saying that if we really want to be attuned to alien life in the cosmos, it's so likely to be in the form of machine intelligence that we ought to "be alert to apparent violations of physics."

These forms of life may well be speaking to us even now. It's just that our radio telescopes, which listen to the skies for signals from alien beings, can't understand what they're hearing. "Even if the search succeeded," Rees writes, "it would still in my view be unlikely that the 'signal' would be a decodable message."

It's a whole new twist on George Berkeley's question. The tree would fall in the forest. We'd hear it. But it would sound nothing like a tree.

Greetings, E.T.
(Please Don't Murder Us.)

BY STEVEN JOHNSON | JUNE 28, 2017

ON NOV. 16, 1974, a few hundred astronomers, government officials and other dignitaries gathered in the tropical forests of Puerto Rico's northwest interior, a four-hour drive from San Juan. The occasion was a rechristening of the Arecibo Observatory, at the time the largest radio telescope in the world. The mammoth structure — an immense concrete-and-aluminum saucer as wide as the Eiffel Tower is tall, planted implausibly inside a limestone sinkhole in the middle of a mountainous jungle — had been upgraded to ensure its ability to survive the volatile hurricane season and to increase its precision tenfold.

To celebrate the reopening, the astronomers who maintained the observatory decided to take the most sensitive device yet constructed for listening to the cosmos and transform it, briefly, into a machine for talking back. After a series of speeches, the assembled crowd sat in silence at the edge of the telescope while the public-address system blasted nearly three minutes of two-tone noise through the muggy afternoon heat. To the listeners, the pattern was indecipherable, but somehow the experience of hearing those two notes oscillating in the air moved many in the crowd to tears.

That 168 seconds of noise, now known as the Arecibo message, was the brainchild of the astronomer Frank Drake, then the director of the organization that oversaw the Arecibo facility. The broadcast marked the first time a human being had intentionally transmitted a message targeting another solar system. The engineers had translated the missive into sound, so that the assembled group would have something to experience during the transmission. But its true medium was the silent, invisible pulse of radio waves, traveling at the speed of light.

It seemed to most of the onlookers to be a hopeful act, if a largely symbolic one: a message in a bottle tossed into the sea of deep space.

But within days, the Royal Astronomer of England, Martin Ryle, released a thunderous condemnation of Drake's stunt. By alerting the cosmos of our existence, Ryle wrote, we were risking catastrophe. Arguing that "any creatures out there [might be] malevolent or hungry," Ryle demanded that the International Astronomical Union denounce Drake's message and explicitly forbid any further communications. It was irresponsible, Ryle fumed, to tinker with interstellar outreach when such gestures, however noble their intentions, might lead to the destruction of all life on earth.

TODAY, MORE THAN four decades later, we still do not know if Ryle's fears were warranted, because the Arecibo message is still eons away from its intended recipient, a cluster of roughly 300,000 stars known as M13. If you find yourself in the Northern Hemisphere this summer on a clear night, locate the Hercules constellation in the sky, 21 stars that form the image of a man, arms outstretched, perhaps kneeling. Imagine hurtling 250 trillion miles toward those stars. Though you would have traveled far outside our solar system, you would only be a tiny fraction of the way to M13. But if you were somehow able to turn on a ham radio receiver and tune it to 2,380 MHz, you might catch the message in flight: a long series of rhythmic pulses, 1,679 of them to be exact, with a clear, repetitive structure that would make them immediately detectable as a product of intelligent life.

In its intended goal of communicating with life-forms outside our planet, the Arecibo message has surprisingly sparse company. Perhaps the most famous is housed aboard the Voyager 1 spacecraft — a gold-plated audiovisual disc, containing multilingual greetings and other evidence of human civilization — which slipped free of our solar system just a few years ago, traveling at a relatively sluggish 35,000 miles per hour. By contrast, at the end of the three-minute transmission of the Arecibo message, its initial pulses had already reached the orbit of Mars. The entire message took less than a day to leave the solar system.

True, some signals emanating from human activity have traveled much farther than even Arecibo, thanks to the incidental leakage of radio and television broadcasts. This was a key plot point in Carl Sagan's novel, "Contact," which imagined an alien civilization detecting the existence of humans through early television broadcasts from the Berlin Olympic Games, including clips of Hitler speaking at the opening ceremony. Those grainy signals of Jesse Owens, and later of Howdy Doody and the McCarthy hearings, have ventured farther into space than the Arecibo pulses. But in the 40 years since Drake transmitted the message, just over a dozen intentional messages have been sent to the stars, most of them stunts of one fashion or another, including one broadcast of the Beatles' "Across the Universe" to commemorate the 40th anniversary of that song's recording. (We can only hope the aliens, if they exist, receive that message before they find the Hitler footage.)

In the age of radio telescopes, scientists have spent far more energy trying to look for signs that other life might exist than they have signaling the existence of our own. Drake himself is now more famous for inaugurating the modern search for extraterrestrial intelligence (SETI) nearly 60 years ago, when he used a telescope in West Virginia to scan two stars for structured radio waves. Today the nonprofit SETI Institute oversees a network of telescopes and computers listening for signs of intelligence in deep space. A new SETI-like project called Breakthrough Listen, funded by a $100 million grant from the Russian billionaire Yuri Milner, promises to radically increase our ability to detect signs of intelligent life. As a species, we are gathered around more interstellar mailboxes than ever before, waiting eagerly for a letter to arrive. But we have, until recently, shown little interest in sending our own.

Now this taciturn phase may be coming to an end, if a growing multidisciplinary group of scientists and amateur space enthusiasts have their way. A newly formed group known as METI (Messaging Extra Terrestrial Intelligence), led by the former SETI scientist

Douglas Vakoch, is planning an ongoing series of messages to begin in 2018. And Milner's Breakthrough Listen endeavor has also promised to support a "Breakthrough Message" companion project, including an open competition to design the messages that we will transmit to the stars. But as messaging schemes proliferate, they have been met with resistance. The intellectual descendants of Martin Ryle include luminaries like Elon Musk and Stephen Hawking, and they caution that an assumption of interstellar friendship is the wrong way to approach the question of extraterrestrial life. They argue that an advanced alien civilization might well respond to our interstellar greetings with the same graciousness that Cortés showed the Aztecs, making silence the more prudent option.

If you believe that these broadcasts have a plausible chance of making contact with an alien intelligence, the choice to send them must rank as one of the most important decisions we will ever make as a species. Are we going to be galactic introverts, huddled behind the door and merely listening for signs of life outside? Or are we going to be extroverts, conversation-starters? And if it's the latter, what should we say?

AMID THE DECOMMISSIONED splendor of Fort Mason, on the northern edge of San Francisco, sits a bar and event space called the Interval. It's run by the Long Now Foundation, an organization founded by Stewart Brand and Brian Eno, among others, to cultivate truly long-term thinking. The group is perhaps most famous for its plan to build a clock that will successfully keep time for 10,000 years. Long Now says the San Francisco space is designed to push the mind away from our attention-sapping present, and this is apparent from the 10,000-year clock prototypes to the menu of "extinct" cocktails.

The Interval seemed like a fitting backdrop for my first meeting with Doug Vakoch, in part because Long Now has been advising METI on its message plans and in part because the whole concept of sending interstellar messages is the epitome of long-term decision-making.

The choice to send a message into space is one that may well not generate a meaningful outcome for a thousand years, or a hundred thousand. It is hard to imagine any decision confronting humanity that has a longer time horizon.

As Vakoch and I settled into a booth, I asked him how he found his way to his current vocation. "I liked science when I was a kid, but I couldn't make up my mind which science," he told me. Eventually, he found out about a burgeoning new field of study known as exobiology, or sometimes astrobiology, that examined the possible forms life could take on other planets. The field was speculative by nature: After all, its researchers had no actual specimens to study. To imagine other forms of life in the universe, exobiologists had to be versed in the astrophysics of stars and planets; the chemical reactions that could capture and store energy in these speculative organisms; the climate science that explains the weather systems on potentially life-compatible planets; the biological forms that might evolve in those different environments. With exobiology, Vakoch realized, he didn't have to settle on one discipline: "When you think about life outside the earth, you get to dabble in all of them."

As early as high school, Vakoch began thinking about how you might communicate with an organism that had evolved on another planet, the animating question of a relatively obscure subfield of exobiology known as exosemiotics. By the time Vakoch reached high school in the 1970s, radio astronomy had advanced far enough to turn exosemiotics from a glorified thought experiment into something slightly more practical. Vakoch did a science-fair project on interstellar languages, and he continued to follow the field during his college years, even as he was studying comparative religion at Carleton College in Minnesota. "The issue that really hit me early on, and that has stayed with me, is just the challenge of creating a message that would be understandable," Vakoch says. Hedging his bets, he pursued a graduate degree in clinical psychology, thinking it might help him better understand the mind of some unknown organism across the universe.

If the exosemiotics passion turned out to be a dead end professionally, he figured that he could always retreat back to a more traditional career path as a psychologist.

During Vakoch's graduate years, SETI was transforming itself from a NASA program sustained by government funding to an independent nonprofit organization, supported in part by the new fortunes of the tech sector. Vakoch moved to California and joined SETI in 1999. In the years that followed, Vakoch and other scientists involved in the program grew increasingly vocal in their argument for sending messages as well as listening for them. The "passive" approach was essential, they argued, but an "active" SETI — one targeting nearby star systems with high-powered radio signals — would increase the odds of contact. Concerned that embracing an active approach would imperil its funding, the SETI board resisted Vakoch's efforts. Eventually Vakoch decided to form his own international organization, METI, with a multidisciplinary team that includes the former NASA chief historian Steven J. Dick, the French science historian Florence Raulin Cerceau, the Indian ecologist Abhik Gupta and the Canadian anthropologist Jerome H. Barkow.

The newfound interest in messaging has been piqued in large part by an explosion of newly discovered planets. We now know that the universe is teeming with planets occupying what exobiologists call "the Goldilocks zone": not too hot and not too cold, with "just right"-surface temperatures capable of supporting liquid water. At the start of Drake's career in the 1950s, not a single planet outside our solar system had been observed. Today we can target a long list of potential Goldilocks-zone planets, not just distant clusters of stars. "Now we know that virtually all stars have planets," Vakoch says, adding that, of these stars, "maybe one out of five have potentially habitable planets. So there's a lot of real estate that could be inhabited."

When Frank Drake and Carl Sagan first began thinking about message construction in the 1960s, their approach was genuinely equivalent to the proverbial message in a bottle. Now, we may not know the exact addresses of planets where life is likely, but we have identified many promising ZIP codes. The recent discovery of the Trappist-1 planets, three of which are potentially habitable, triggered such excitement in part because those planets were, relatively speaking, so close to home: just 40 light-years from Earth. If the Arecibo message does somehow find its way to an advanced civilization in M13, word would not come back for at least 50,000 years. But a targeted message sent to Trappist-1 could generate a reply before the end of the century.

FRANK DRAKE IS NOW 87 and lives with his wife in a house nestled in an old-growth redwood forest, at the end of a narrow, winding road in the hills near Santa Cruz. His circular driveway wraps around the trunk of a redwood bigger than a pool table. As I left my car, I found myself thinking again of the long now: a man who sends messages with a potential life span of 50,000 years, living among trees that first took root a millennium ago.

Drake has been retired for more than a decade, but when I asked him about the Arecibo message, his face lit up at the memory. "We had just finished a very big construction project at Arecibo, and I was director then, and so they said, 'Can you please arrange a big ceremony?'" he recalled. "We had to have some kind of eye-catching event for this ceremony. What could we do that would be spectacular? We could send a message!"

But how can you send a message to a life-form that may or may not exist and that you know nothing at all about, other than the fact that it evolved somewhere in the Milky Way? You need to start by explaining how the message is supposed to be read, which is known in exosemiotics as the "primer." You don't need a primer on Earth: You point to a cow, and you say, "Cow." The plaques that NASA sent into space with Pioneer and Voyager had the advantage of being physical objects that could convey visual information, which at least enables you to connect words with images of the objects they refer to. In other words, you draw a cow and then put the word "cow" next to the drawing and slowly, with enough pointing, a language comes into view. But physical objects can't be moved fast enough to get to a potential recipient in useful time scales. You need electromagnetic waves if you want to reach across the Milky Way.

But how do you point to something with a radio wave? Even if you figured out a way to somehow point to a cow with electromagnetic signals, the aliens aren't going to have cows in their world, which means the reference will most likely be lost on them. Instead, you need to think hard about the things that our hypothetical friends in the Trappist-1 system will have in common with us. If their civilization is advanced enough to recognize structured data in radio waves, they must share many of our scientific and technological concepts. If they are hearing our message, that means they are capable of parsing structured disturbances in the electromagnetic spectrum, which means they understand the electromagnetic spectrum in some meaningful way.

The trick, then, is just getting the conversation started. Drake figured that he could count on intelligent aliens possessing the concept of simple numbers: one, three, 10, etc. And if they have numbers, then they will also very likely have the rest of what we know as basic math: addition, subtraction, multiplication, division. Furthermore, Drake reasoned, if they have multiplication and division, then they are likely to understand the concept of prime numbers — the group of numbers that are divisible only by themselves and one. (In "Contact," the intercepted alien message begins with a long string of primes: 1, 2, 3, 5, 7, 11, 13, 17, 19, 23, and so on.) Many objects in space, like pulsars, send out radio signals with a certain periodicity: flashes of electromagnetic activity that switch on and off at regular rates. Primes, however, are a telltale sign of intelligent life. "Nature never uses prime numbers," Drake says. "But mathematicians do."

Drake's Arecibo message drew upon a close relative of the prime numbers to construct its message. He chose to send exactly 1,679 pulses, because 1,679 is a semiprime number: a number that can be formed only by multiplying two prime numbers together, in this case 73 and 23. Drake used that mathematical quirk to turn his pulses of electromagnetic energy into a visual system. To simplify his approach, imagine I send you a message consisting of 10 X's and 5 O's: XOXOXXXXOXXOXOX. You notice that the number 15 is a semiprime number, and so you organize the symbols in a 3-by-5 grid and leave the O's as blank spaces. The result is this:

If you were an English speaker, you might just recognize a greeting in that message, the word "HI" mapped out using only a binary language of on-and-off states.

Drake took the same approach, only using a much larger semiprime, which gave him a 23-by-73 grid to send a more complicated message. Because his imagined correspondents in M13 were not likely to understand any human language, he filled the grid with a mix of mathematical and visual referents. The top of the grid counted from one to 10 in binary code — effectively announcing to the aliens that numbers will be represented using these symbols.

Having established a way of counting, Drake then moved to connect the concept of numbers to some reference that the citizens of M13 would likely share with us. For this step, he encoded the atomic numbers for five elements: hydrogen, carbon, nitrogen, oxygen and phosphorous, the building blocks of DNA. Other parts of the message were more visually oriented. Drake used the on-off pulses of the radio signal to "draw" a pixelated image of a human body. He also included a sketch of our solar system and of the Arecibo telescope itself. The message said, in effect: This is how we count; this is what we are made of; this is where we came from; this is what we look like; and this is the technology we are using to send this message to you.

AS INVENTIVE AS DRAKE'S exosemiotics were in 1974, the Arecibo message was ultimately more of a proof-of-concept than a genuine attempt to make contact, as Drake himself is the first to admit. For starters, the 25,000 light-years that separate us from M13 raise a legitimate question about whether humans will even be around — or recognizably human — by the time a message comes back. The choice of where to send it was almost entirely haphazard. The METI project intends to improve on the Arecibo model by directly targeting nearby Goldilocks-zone planets.

One of the most recent planets added to that list orbits the star Gliese 411, a red dwarf located eight light-years away from Earth. On a spring evening in the Oakland hills, our own sun putting on a spectacular display as it slowly set over the Golden Gate Bridge, Vakoch and I met at one of the observatories at the Chabot Space and Science Center to take a peek at Gliese 411. A half moon overhead reduced our visibility but not so much that I couldn't make out the faint tangerine glimmer of the star, a single blurred point of light that had traveled nearly 50 trillion miles across the universe to land on my retina. Even with the power of the Oakland telescope, there was no way to spot a planet orbiting the red dwarf. But in February of this year, a team of researchers using the Keck I telescope at the top of Mauna Kea in Hawaii announced that they had detected a "super-earth" in orbit around Gliese, a rocky and hot planet larger than our own.

The METI group aims to improve on the Arecibo message not just by targeting specific planets, like that super-earth orbiting Gliese, but also by rethinking the nature of the message itself. "Drake's original design plays into the bias that vision is universal among intelligent life," Vakoch told me. Visual diagrams — whether formed through semiprime grids or engraved on plaques — seem like a compelling way to encode information to us because humans happen to have evolved an unusually acute sense of vision. But perhaps the aliens followed a different evolutionary path and found their way to a technologically advanced civilization with an intelligence that was rooted in some other sense: hearing, for example, or some other way of perceiving the world around them for which there is no earthly equivalent.

Like so much of the SETI/METI debate, the question of visual messaging quickly spirals out into a deeper meditation, in this instance on the connection between intelligence and visual acuity. It is no accident that eyes developed independently so many times over the course of evolution on Earth, given the fact that light conveys information faster than any other conduit. That transmission-speed advantage would presumably apply on other planets in the Goldilocks zone, even if they happened to be on the other side of the Milky Way, and so it seems plausible that intelligent creatures would evolve some sort of visual system as well.

But even more universal than sight would be the experience of time. Hans Freudenthal's "Lincos: Design of a Language for Cosmic Intercourse," a seminal book of exosemiotics published more than a half-century ago, relied heavily on temporal cues in its primer stage. Vakoch and his collaborators have been working with Freudenthal's language in their early drafts for the message. In Lincos, duration is used as a key building block. A pulse that lasts for a certain stretch (say, in human terms, one second) is followed by a sequence of pulses that signify the "word" for one; a pulse that lasts for six seconds is followed by the word for six. The words for basic math properties can be conveyed by combining pulses of different lengths. You might demonstrate the property of addition by sending the word for "three" and "six" and then sending a pulse that lasts for nine seconds. "It's a way of being able to point at objects when you don't have anything right in front of you," Vakoch explains.

Other messaging enthusiasts think we needn't bother worrying about primers and common referents. "Forget about sending mathematical relationships, the value of pi, prime numbers or the Fibonacci series," the senior SETI astronomer, Seth Shostak, argued in a 2009 book. "No, if we want to broadcast a message from Earth, I propose that we just feed the Google servers into the transmitter. Send the aliens the World Wide Web. It would take half a year or less to transmit this in the microwave; using infrared lasers shortens the transmit time to no more than two days." Shostak believes that the sheer magnitude of the transmitted data would enable the aliens to decipher it. There is some precedent for this in the history of archaeologists studying dead languages: The hardest code to crack is one with only a few fragments.

Sending all of Google would be a logical continuation of Drake's 1974 message, in terms of content if not encoding. "The thing about the Arecibo message is that, in a sense, it's brief but its intent is encyclopedic," Vakoch told me as we waited for the sky to darken in the Oakland hills. "One of the things that we are exploring for our transmission is the opposite extreme. Rather than being encyclopedic, being selective. Instead of this huge digital data dive, trying to do something elegant. Part of that is thinking about what are the most fundamental concepts we need." There is something provocative about the question Vakoch is wrestling with here: Of all the many manifestations of our achievements as a species, what's the simplest message we can create that will signal that we're interesting, worthy of an interstellar reply?

But to METI's critics, what he should be worrying about instead is the form that the reply might take: a death ray, or an occupying army.

BEFORE DOUG VAKOCH had even filed the papers to form the METI nonprofit organization in July 2015, a dozen or so science-and-tech luminaries, including SpaceX's Elon Musk, signed a statement categorically opposing the project, at least without extensive further discussion, on a planetary scale. "Intentionally signaling other civilizations in the Milky Way Galaxy," the statement argued, "raises concerns from all the people of Earth, about both the message and the consequences of contact. A worldwide scientific, political and humanitarian discussion must occur before any message is sent."

One signatory to that statement was the astronomer and science-fiction author David Brin, who has been carrying on a spirited but collegial series of debates with Vakoch over the wisdom of his project. "I just don't think anybody should give our children a fait accompli based on blithe assumptions and assertions that have been untested and not subjected to critical peer review," he told me over a Skype call from his home office in Southern California. "If you are going to do something that is going to change some of the fundamental observable parameters of our solar system, then how about an environmental-impact statement?"

The anti-METI movement is predicated on a grim statistical likelihood: If we do ever manage to make contact with another intelligent life-form, then almost by definition, our new pen pals will be far more advanced than we are. The best way to understand this is to consider, on a percentage basis, just how young our own high-tech civilization actually is. We have been sending structured radio signals from Earth for only the last 100 years. If the universe were exactly 14 billion years old, then it would have taken 13,999,999,900 years for radio communication to be harnessed on our planet. The odds that our message would reach a society that had been tinkering with radio for a shorter, or even similar, period of time would be staggeringly long. Imagine another planet that deviates from our timetable by just a tenth of 1 percent: If they are more advanced than us, then they will have been using radio (and successor technologies) for 14 million years. Of course, depending on where they live in the universe, their signals might take millions of years to reach us. But even if you factor in that transmission lag, if we pick up a signal from another galaxy, we will almost certainly find ourselves in conversation with a more advanced civilization.

It is this asymmetry that has convinced so many future-minded thinkers that METI is a bad idea. The history of colonialism here on Earth weighs particularly heavy on the imaginations of the METI critics. Stephen Hawking, for instance, made this observation in a 2010 documentary series: "If aliens visit us, the outcome would be much as when Columbus landed in America, which didn't turn out well for the Native Americans." David Brin echoes the

Hawking critique: "Every single case we know of a more technologically advanced culture contacting a less technologically advanced culture resulted at least in pain."

METI proponents counter the critics with two main arguments. The first is essentially that the horse has already left the barn: Given that we have been "leaking" radio waves in the form of "Leave It to Beaver" and the nightly news for decades, and given that other civilizations are likely to be far more advanced than we are, and thus capable of detecting even weak signals, then it seems likely that we are already visible to extraterrestrials. In other words, they know we're here, but they haven't considered us to be worthy of conversation yet. "Maybe in fact there are a lot more civilizations out there, and even nearby planets are populated, but they're simply observing us," Vakoch argues. "It's as if we are in some galactic zoo, and if they've been watching us, it's like watching zebras talking to one another. But what if one of those zebras suddenly turns toward you and with its hooves starts scratching out the prime numbers. You'd relate to that zebra differently!"

Brin thinks that argument dangerously underestimates the difference between a high-power, targeted METI transmission and the passive leakage of media signals, which are far more difficult to detect. "Think about it this way: If you want to communicate with a Boy Scout camp on the other side of the lake, you could kneel down at the end of the lake and slap the water in Morse code," he says. "And if they are spectacularly technologically advanced Boy Scouts who happened also to be looking your way, they might build instruments that would be able to parse out your Morse code. But then you whip out your laser-pointer and point it at their dock. That is exactly the order of magnitude difference between picking up [reruns of] 'I Love Lucy' from the 1980s, when we were at our noisiest, and what these guys want to do."

METI defenders also argue that the threat of some Klingon-style invasion is implausible, given the distances involved. If, in fact, advanced civilizations were capable of zipping around the galaxy at the speed of light, we would have already encountered them. The much more likely situation is that only communications can travel that fast, and so a malevolent presence on some distant planet will only be able to send us hate mail. But critics think that sense of security is unwarranted. Writing in Scientific American, the former chairman of SETI, John Gertz, argued that "a civilization with malign intent that is only modestly more advanced than we are might be able to annihilate Earth with ease by means of a small projectile filled with a self-replicating toxin or nano gray goo; a kinetic missile traveling at an appreciable percentage of the speed of light; or weaponry beyond our imagination."

Brin looks to our own technological progress as a sign of where a more advanced civilization might be in terms of interstellar combat: "It is possible that within just 50 years, we could create an antimatter rocket that could propel a substantial pellet of several kilograms, at half the speed of light at times to intersect with the orbit of a planet within 10 light-years of us." Even a few kilograms colliding at that speed would produce an explosion much greater than the Hiroshima and Nagasaki detonations combined. "And if we could do that in 50 years, imagine what anybody else could do, completely obeying Einstein and the laws of physics."

Interestingly, Frank Drake himself is not a supporter of the METI efforts, though he does not share Hawking and Musk's fear of interstellar conquistadors. "We send messages all the time, free of charge," he says. "There's a big shell out there now 80 light-years around us. A civilization only a little more advanced than we are can pick those things up. So the point is we are already sending copious amounts of information." Drake believes that any other advanced civilization out there must be doing the same, so scientists like Vakoch should devote themselves to picking up on that chatter instead of trying to talk back. METI will consume resources, Drake says, that would be "better spent listening and not sending."

METI CRITICS, OF COURSE, might be right about the frightening sophistication of these other, presumably older civilizations but wrong about the likely nature of their response. Yes, they could be capable of sending projectiles across the galaxy at a quarter of the speed of light. But their longevity would also suggest that they have figured out how to avoid self-destruction on a planetary scale. As Steven Pinker has argued, human beings have become steadily less violent over the last 500 years; per capita deaths from military conflict are most likely at an all-time low. Could this be a recurring pattern throughout the universe, played out on much longer time scales: the older a civilization gets, the less warlike it becomes? In which case, if we do get a message to extraterrestrials, then perhaps they really will come in peace.

These sorts of questions inevitably circle back to the two foundational thought experiments that SETI and METI are predicated upon: the Fermi Paradox and the Drake Equation. The paradox, first formulated by the Italian physicist and Nobel laureate Enrico Fermi, begins with the assumption that the universe contains an unthinkably large number of stars, with a significant percentage of them orbited by planets in the Goldilocks zone. If intelligent life arises on even a small fraction of those planets, then the universe should be teeming with advanced civilizations. And yet to date, we have seen no evidence of those civilizations, even after several decades of scanning the skies through SETI searches. Fermi's question, apparently raised during a lunch conversation at Los Alamos in the early 1950s, was a simple one: "Where is everybody?"

The Drake Equation is an attempt to answer that question. The equation dates back to one of the great academic retreats in the history of scholarship: a 1961 meeting at the Green Bank observatory in West Virginia, which included Frank Drake, a 26-year-old Carl Sagan and the dolphin researcher (and later psychedelic explorer) John Lilly. During the session, Drake shared his musings on the Fermi Paradox, formulated as an equation. If we start scanning the cosmos for signs of intelligent life, Drake asked, how likely are we to actually detect something? The equation didn't generate a clear answer, because almost all the variables were unknown at the time and continue to be largely unknown a half-century later. But the equation had a clarifying effect, nonetheless. In mathematical form, it looks like this:

$$N = R^* \times fp \times ne \times fl \times fi \times fc \times L$$

N represents the number of extant, communicative civilizations in the Milky Way. The initial variable R^* corresponds to the rate of star formation in the galaxy, effectively giving you the total number of potential suns that could support life. The remaining variables then serve as a kind of nested sequence of filters: Given the number of stars in the Milky Way, what fraction of those have planets, and how many of those have an environment that can support life? On those potentially hospitable planets, how often does life itself actually emerge, and what fraction of that life evolves into intelligent life, and what fraction of that life eventually leads to a civilization's transmitting detectable signals into space? At the end of his equation, Drake placed the crucial variable L, which is the average length of time during which those civilizations emit those signals.

What makes the Drake Equation so mesmerizing is in part the way it forces the mind to yoke together so many different intellectual disciplines in a single framework. As you move from left to right in the equation, you shift from astrophysics, to the biochemistry of life, to evolutionary theory, to cognitive science, all the way to theories of technological development. Your guess about each value in the Drake Equation winds up revealing a whole worldview: Perhaps you think life is rare, but when it does emerge, intelligent life usually follows; or perhaps you think microbial life is ubiquitous throughout the cosmos, but more complex organisms almost never form. The equation is notoriously vulnerable to very different outcomes, depending on the numbers you assign to each variable.

The most provocative value is the last one: L, the average life span of a signal-transmitting civilization. You don't have to be a Pollyanna to defend a relatively high L value. All you need is to believe that it is possible for civilizations to become fundamentally self-sustaining and survive for millions of years. Even if one in a thousand intelligent life-forms in space generates a million-year civilization, the value of L increases meaningfully. But if your L-value is low, that implies a further question: What is keeping it low? Do technological civilizations keep flickering on and off in the Milky Way, like so many fireflies in space? Do they run out of resources? Do they blow themselves up?

Since Drake first sketched out the equation in 1961, two fundamental developments have reshaped our understanding of the problem. First, the numbers on the left-hand side of the equation (representing the amount of stars with habitable planets) have increased by several orders of magnitude. And second, we have been listening for signals for decades and heard nothing. As Brin puts it: "Something is keeping the Drake Equation small. And the difference between all the people in the SETI debates is not whether that's true, but where in the Drake panoply the fault lies."

If the left-hand values keep getting bigger and bigger, the question is which variables on the right-hand side are the filters. As Brin puts it, we want the filter to be behind us, not the one variable, L, that still lies ahead of us. We want the emergence of intelligent life to be astonishingly rare; if the opposite is true, and intelligent life is abundant in the Milky Way, then L values might be low, perhaps measured in centuries and not even millenniums. In that case, the adoption of a technologically advanced lifestyle might be effectively simultaneous with extinction. First you invent radio, then you invent technologies capable of destroying all life on your planet and shortly thereafter you push the button and your civilization goes dark.

The L-value question explains why so many of METI's opponents — like Musk and Hawking — are also concerned with the threat of extinction-level events triggered by other potential threats: superintelligent computers, runaway nanobots, nuclear weapons, asteroids. In a low L-value universe, planet-wide annihilation is an imminent possibility. Even if a small fraction of alien civilizations out there would be inclined to shoot a two-kilogram pellet toward us at half the speed of light, is it worth sending a message if there's even the slightest chance that the reply could result in the destruction of all life on earth?

Other, more benign, explanations for the Fermi Paradox exist. Drake himself is pessimistic about the L value, but not for dystopian reasons. "It's because we're getting better at technology," he says. The modern descendants of the TV and radio towers that inadvertently sent Elvis to the stars are far more efficient in terms of the power they use, which means the "leaked" signals emanating from Earth are far fainter than they were in the 1950s. In fact, we increasingly share information via fiber optics and other terrestrial conduits that have zero leakage outside our atmosphere. Perhaps technologically advanced societies do flicker on and off like fireflies, but it's not a sign that they're self-destructive; it's just a sign that they got cable.

But to some METI critics, even a less-apocalyptic interpretation of the Fermi Paradox still suggests caution. Perhaps advanced civilizations tend to reach a point at which they decide, for some unknown reason, that it is in their collective best interest not to transmit any detectable signal to their neighbors in the Milky Way. "That's the other answer for the Fermi Paradox," Vakoch says with a smile. "There's a Stephen Hawking on every planet, and that's why we don't hear from them."

IN HIS CALIFORNIA home among the redwoods, Frank Drake has a version of the Arecibo message visually encoded in a very different format: not a series of radio-wave pulses but as a stained-glass window in his living room. A grid of pixels on a cerulean blue background, it almost resembles a game of Space Invaders. Stained glass is an appropriate medium, given the nature of the message: an offering dispatched to unknown beings residing somewhere in the sky.

There is something about the METI question that forces the mind to stretch beyond its usual limits. You have to imagine some radically different form of intelligence, using only your human intelligence. You have to imagine time scales on which a decision made in 2017 might trigger momentous consequences 10,000 years from now. The sheer magnitude of those consequences challenges our usual measures of cause and effect. Whether you believe that the aliens are likely to be warriors or Zen masters, if you think that METI has a reasonable chance of making contact with another intelligent organism somewhere in the Milky Way, then you have to accept that this small group of astronomers and science-fiction authors and billionaire patrons debating semiprime numbers and the ubiquity of visual intelligence may in fact be wrestling with a decision that could prove to be the most transformative one in the history of human civilization.

All of which takes us back to a much more down-to-earth, but no less challenging, question: Who gets to decide? After many years of debate, the SETI community established an agreed-upon procedure that scientists and government agencies should follow in the event that the SETI searches actually stumble upon an intelligible signal from space. The protocols specifically ordain that "no response to a signal or other evidence of extraterrestrial intelligence should be sent until appropriate international consultations have taken place." But an equivalent set of guidelines does not yet exist to govern our own interstellar outreach.

One of the most thoughtful participants in the METI debate, Kathryn Denning, an anthropologist at York University in Toronto, has argued that our decisions about extraterrestrial contact are ultimately more political than scientific. "If I had to take a position, I'd say that broad consultation regarding METI is essential, and so I greatly respect the efforts in that direction," Denning says. "But no matter how much consultation there is, it's inevitable that there will be significant disagreement about the advisability of transmitting, and I don't think this is the sort of thing where a simple majority vote or even supermajority should carry the day ... so this keeps bringing us back to the same key question: Is it O.K. for some people to transmit messages at significant power when other people don't want them to?"

In a sense, the METI debate runs parallel to other existential decisions that we will be confronting in the coming decades, as our technological and scientific powers increase. Should we create superintelligent machines that exceed our own intellectual capabilities by such a wide margin that we cease to understand how their intelligence works? Should we "cure" death, as many technologists are proposing? Like METI, these are potentially among the most momentous decisions human beings will ever make, and yet the number of people actively participating in those decisions — or even aware such decisions are being made — is minuscule.

"I think we need to rethink the message process so that we are sending a series of increasingly inclusive messages," Vakoch says. "Any message that we initially send would be too narrow, too incomplete. But that's O.K. Instead, what we should be doing is thinking about how to make the next round of messages better and more inclusive. We ideally want a way to incorporate both technical expertise — people who have been thinking about these issues from a range of different disciplines — and also getting lay input. I think it's often been one or the other. One way we can get lay input in a way that makes a difference in terms of message content is to survey people about what sorts of things they would want to say. It's important to see what the general themes are that people would want to say and then translate those into a Lincos-like message."

When I asked Denning where she stands on the METI issue, she told me: "I have to answer that question with a question: Why are you asking me? Why should my opinion matter more than that of a 6-year-old girl in Namibia? We both have exactly the same amount at stake, arguably, she more than I, since the odds of being dead before any consequences of transmission occur are probably a bit higher for me, assuming she has access to clean water and decent health care and isn't killed far too young in war." She continued: "I think the METI debate may be one of those rare topics where scientific knowledge is highly relevant to the discussion, but its connection to obvious policy is tenuous at best, because in the final analysis, it's all about how much risk the people of Earth are willing to tolerate. ... And why exactly should astronomers, cosmologists, physicists, anthropologists, psychologists, sociologists, biologists, sci-fi authors or anyone else (in no particular order), get to decide what those tolerances should be?"

Wrestling with the METI question suggests, to me at least, that the one invention human society needs is more conceptual than technological: We need to define a special class of decisions that potentially create extinction-level risk. New technologies (like superintelligent computers) or interventions (like METI) that pose even the slightest risk of causing human extinction would require some novel form of global oversight. And part of that process would entail establishing, as Denning suggests, some measure of risk tolerance on a planetary level. If we don't, then by default the gamblers will always set the agenda, and the rest of us will have to live with the consequences of their wagers.

In 2017, the idea of global oversight on any issue, however existential the threat it poses, may sound naïve. It may also be that technologies have their own inevitability, and we can only rein them in for so long: If contact with aliens is technically possible, then someone, somewhere is going to do it soon enough. There is not a lot of historical precedent for humans voluntarily swearing off a new technological capability — or choosing not to make contact with another society — because of some threat that might not arrive for generations. But maybe it's time that humans learned how to make that kind of choice. This turns out to be one of the surprising gifts of the METI debate, whichever side you happen to take. Thinking hard about what kinds of civilization we might be able to talk to ends up making us think even harder about what kind of civilization we want to be ourselves.

Near the end of my conversation with Frank Drake, I came back to the question of our increasingly quiet planet: all those inefficient radio and television signals giving way to the undetectable transmissions of the internet age. Maybe that's the long-term argument for sending intentional messages, I suggested; even if it fails in our lifetime, we will have created a signal that might enable an interstellar connection thousands of years from now.

Drake leaned forward, nodding. "It raises a very interesting, nonscientific question, which is: Are extraterrestrial civilizations altruistic? Do they recognize this problem and establish a beacon for the benefit of the other folks out there? My answer is: I think it's actually Darwinian; I think evolution favors altruistic societies. So my guess is yes. And that means there might be one powerful signal for each civilization." Given the transit time across the universe, that signal might well outlast us as a species, in which case it might ultimately serve as a memorial as much as a message, like an interstellar version of the Great Pyramids: proof that a technologically advanced organism evolved on this planet, whatever that organism's ultimate fate.

As I stared at Drake's stained-glass Arecibo message, in the middle of that redwood grove, it seemed to me that an altruistic civilization — one that wanted to reach across the cosmos in peace — would be something to aspire to, despite the potential for risk. Do we want to be the sort of civilization that boards up the windows and pretends that no one is home, for fear of some unknown threat lurking in the dark sky? Or do we want to be a beacon?

STEVEN JOHNSON is the author of 10 books, most recently "Wonderland: How Play Made the Modern World."

An Interstellar Visitor Both Familiar and Alien

COLUMN | BY DENNIS OVERBYE | NOV. 22, 2017

VISIT THE GALAXY before the galaxy visits you.

This fall, the galaxy came calling in the form of a small reddish cigar-shaped object named Oumuamua by astronomers based in Hawaii. They discovered it in October, careening through the solar system at 40,000 miles an hour, an interstellar emissary from points unknown.

Oumuamua (Oh-moo-a-moo-a), Hawaiian for "scout" or "messenger," was not here long.

It was first noticed zooming out of the constellation Lyra on Oct. 19, about 20 million miles from Earth. By next May, it will already be passing Jupiter on its way out of the solar system.

The asteroid brought shades of Arthur C. Clarke's novel "Rendezvous With Rama," in which explorers find and board an empty alien spacecraft sailing through the solar system. Or perhaps even reminders of the monoliths that power human evolution in "2001: A Space Odyssey."

The discovery set off a worldwide scramble for telescope time to observe the object. Astronomers from the SETI Institute even got into the act, swinging into action to look for alien radio signals Just In Case.

For now, however, those are just science fiction thrills. "Our observations are entirely consistent with it being a natural object," said Karen Meech of the University of Hawaii's Institute for Astronomy and leader of the international collaboration that discovered Oumuamua with the Pan-STARRS 1 telescope on Haleakala, Maui.

Dr. Meech's team has now published the first report of their observations in Nature. The paper describes the interstellar visitor as both reassuringly familiar and utterly alien.

"We don't see anything like that in our solar system," Dr. Meech said.

EUROPEAN SOUTHERN OBSERVATORY/M. KORNMESSER/NASA

Since the asteroid named Oumuamua was first noticed flying through our solar system in October, researchers have been monitoring for alien signals, so far to no avail.

In its color and other imputed properties, Oumuamua resembles the asteroids we already know and fear will one day smash the Earth and human civilization to smithereens.

But the asteroid's shape is weird. It is extremely elongated, at least 10 times as long as it is wide, perhaps 800 yards by 80 yards.

Though the mysterious object is nearly gone, thousands like it probably lurk unsuspected and undetected in our solar system, according to the scientists.

The Pan-STARRS telescope was built to patrol the sky for dangerous asteroids in our own system, not interlopers from beyond. But astronomers got a surprise.

Dr. Meech learned in a phone call one night that her colleagues had found one whose path seemed to originate beyond the solar system altogether. "Wow, this is exciting," Dr. Meech recalled thinking.

Astronomers had long surmised that interstellar debris might invade the solar system from time to time, in the form of icy chunks spit from the rocky disks forming faraway planets.

Such wanderers would manifest themselves as comets when they got close to our sun, vaporizing and lighting up; however, they have not been seen. Now astronomers know why.

Oumuamua showed no such cometary brightening. It is so dark and faint that it could only have been detected by a powerful telescope with a wide field of view, like Pan-STARRS.

Many more should be visible to the Large Synoptic Survey Telescope, with a diameter of eight meters, being built in Chile. "We have to get ready for these," Dr. Meech said.

Oumuamua brightens and dims dramatically every 7.3 hours, which suggests that it is rotating about its short axis. That is something the little asteroid could endure without flying apart only if it were made of sterner, stronger stuff than the dirty snow that characterizes most comets.

Spectral measurements have revealed that Oumuamua is dark red, the color of many moons of the outer solar system on which icy organic molecules have been stained by radiation in outer space. Iron can also contribute that color, Dr. Meech said.

How Oumuamua got its shape is a mystery for now. Perhaps, Dr. Meech said, it was shot away from its home star by a supernova explosion. Or perhaps it was formed by a pair of objects that collided and stuck together. Stay tuned.

Where did it come from? Dr. Meech said the astronomers were initially excited when the orbit appeared to point to the brightest star in Lyra, Vega, which is known to have a debris disk. It would have taken the object about 600,000 years to get here from there, astronomers estimated.

But further refinements in the trajectory have made it less likely that Vega actually was the source.

The fact that Oumuamua was traveling at about the same speed relative to the sun as other nearby stars suggests that this is the asteroid's first encounter with a new star system.

Still, the authors write in Nature, "The possibility that Oumuamua has been orbiting the galaxy for billions of years cannot be ruled out."

Where it's going is equally in the dark. Like the Voyager spacecraft slingshotted around Jupiter, Oumuamua will leave the sun with more energy and heading in a different direction, Dr. Meech said.

The adventures of this asteroid and its ilk paint a very different picture of the galaxy than you might imagine while gazing up at a sky in which the stars seem separate and sovereign, beaming away in solitude.

The oxygen and iron in our blood were created in a supernova explosion long ago and far away from here, and the gold in our wedding bands was formed in the collision of neutron stars. We now know that meteorites sprung by asteroid impacts on Mars land on Earth all the time.

Otherwise respectable astronomers speculate that one of them might have seeded Earth with life that started on Mars when it was warm and wet long ago.

But we can look even further out and backward in time for our connection to the cosmos. Consider the hundreds of thousands of years that Oumuamua might have taken to get here. While that might sound like a long time, it is a blink of cosmic time.

The Milky Way galaxy is 10 billion years old. Which means that over the course of our galaxy's lifetime so far, little Oumuamua might have cruised through some 20,000 star systems — a small fraction of the 200 billion stars in the galaxy, but still a goodly number of stamps on its cosmic passport.

Oumuamua would have trailed behind bits of dust and debris, and so the stars and the worlds of the galaxy mix it up. It may be that the universe is a small place after all.

DENNIS OVERBYE writes the Out There column for The New York Times.

A Nearby Earth-Size Planet May Have Conditions for Life

BY KENNETH CHANG | NOV. 15, 2017

THERE'S A NEW PLACE to look for life in the universe.

Astronomers announced on Wednesday the discovery of an Earth-size planet around a small red star in our corner of the galaxy. The planet could hold liquid water and conditions favorable for life.

The star, Ross 128, is not the closest with a planet similar in size to ours. That would be the sun's next door neighbor, Proxima Centauri, just 4.2 light-years away.

And there appears to be just one planet orbiting Ross 128 — not the bounty of seven Earth-size planets that circle Trappist-1, a red dwarf about 40 light-years from here.

But unlike those stars, Ross 128, about 11 light-years from Earth, appears to be a quiet, well-behaved star, without the violent eruptions of radiation that might wipe out any beginnings of life before they had a chance to take hold on the planet.

"Those flares can sterilize the atmosphere of the planet," said Xavier Bonfils of the Institute of Planetology and Astrophysics in Grenoble, France, the lead author of a paper describing the planet. "Ross 128 is one of the quietest stars of the neighborhood."

The findings appear in the journal Astronomy and Astrophysics.

The astronomers did not directly see the planet but instead used a telescope in Chile to measure wobbles in the wavelengths of light coming from the star. The wobbles are caused by the gravitational pull of the unseen planet.

The magnitude of the wobbles indicates that the planet is at least 1.35 times the mass of Earth but could easily be twice the mass of Earth.

Astronomers' instruments are not yet sensitive enough to spot Earth-size planets in Earthlike orbits around stars similar to our sun. It is easier to detect Earth-size planets around dimmer and cooler stars known as red dwarfs, which are the most common type of star in the Milky Way.

Astronomers have in the past couple of decades discovered an abundance of star-hugging planets, far different from anything in our solar system. The Ross 128 planet is only about 4.5 million miles from the star, much closer than the 93 million miles between Earth and the sun. Even Mercury, the innermost planet of the solar system, is 36 million miles from the sun.

If the newly discovered planet were the same distance from the red dwarf as Earth is from the sun, it would be frigid. But it is close enough to Ross 128 that it absorbs warmth sufficient for liquid water, one of the requisite ingredients for life, to potentially exist on the surface. (If anything, the planet may be too warm, more like the planet Venus.)

Dr. Bonfils said Ross 128 appears to be at least five billion years old — older than our solar system — and perhaps as old as 10 billion years. The star may have been more turbulent in its youth. But even if solar flares billions of years ago stripped away the planet's atmosphere, it could have been replenished by gases emanating from the planet's interior, Dr. Bonfils said.

Vladimir Airapetian, an astrophysicist at the NASA Goddard Space Flight Center in Greenbelt, Md., questioned whether Ross 128 would be such a benign star.

"Even being quiet, its X-ray to extreme U.V. emission can be 10 times higher than that of the sun," said Dr. Airapetian. That amount of radiation might be enough to destroy the planet's atmosphere.

In an Astrophysical Journal Letters article in February, he and his colleagues noted that radiation from red dwarf stars might strip oxygen from the atmospheres of nearby planets.

William C. Danchi, also a Goddard astrophysicist and an author of that article, was more positive.

"There is potential for an atmosphere and hence habitability, but it is highly uncertain," he said. "This is an important discovery and well worth many follow-up studies."

The next generation of large terrestrial telescopes, with mirrors 100 feet or more in diameter, should be able to make out the planet circling Ross 128 and possibly identify specific molecules in its atmosphere.

"It would be rather easy to search for oxygen in the atmosphere of such a planet," Dr. Bonfils said.

A Large Body of Water on Mars Is Detected, Raising the Potential for Alien Life

BY KENNETH CHANG AND DENNIS OVERBYE | JULY 25, 2018

The discovery suggests that the liquid conditions beneath the icy southern polar cap may have provided one of the critical building blocks for life on the red planet.

FOR THE FIRST TIME, scientists have found a large, watery lake beneath an ice cap on Mars. Because water is essential to life, the discovery offers an exciting new place to search for life-forms beyond Earth.

Italian scientists working on the European Space Agency's Mars Express mission announced on Wednesday that a 12-mile-wide underground liquid pool — not just the momentary damp spots seen in the past — had been detected by radar measurements near the Martian south pole.

"Water is there," Enrico Flamini, the former chief scientist of the Italian Space Agency who oversaw the research, said during a news conference.

"It is liquid, and it's salty, and it's in contact with rocks," he added. "There are all the ingredients for thinking that life can be there, or can be maintained there if life once existed on Mars."

The body of water appears similar to underground lakes found on Earth in Greenland and Antarctica. On Earth, microbial life persists down in the dark, frigid waters of one such lake. The ice on Mars would also shield the Martian lake from the damaging radiation that bombards the planet's surface.

Jonathan Lunine, director of the Center for Astrophysics and Planetary Science at Cornell University, who was not involved with the research, said the finding transforms Mars from a dusty planet to yet another "ocean world" in the solar system.

NASA

NASA's Mars Reconnaissance Orbiter observed this surface texture in 2017, which is the result of rock interacting with water. The conditions of the interaction are unclear.

"I think the more we explore Mars, the more intriguing and complex it becomes," Dr. Lunine said.

For years, "follow the water" has been the mantra of NASA and indeed humanity's search for life somewhere else. Without water, there is no life as we know it. In recent years, that has led the space agency to contemplate robot probes to the moons of Jupiter and Saturn, like Europa or Enceladus, where it is now known that salty oceans exist underneath thin shells of ice and where imaginative astrobiologists can envision microbes or more complex creatures.

Since humans could see through telescopes across space, Mars has been the favorite abode of imaginary life, the backyard just over the fence where the astronomer Percival Lowell imagined he could see canals and even cities webbing the orange globe. In the final evenings of this month, the planet looms like a red lantern in the East, just 35,784,871 miles from Earth — the closest it has been in 15 years.

Those early science fiction visions were dashed when the first spacecraft photos of the planet revealed a dry, cratered and lifeless-looking surface — a seemingly dead planet. In the history of Mars exploration ever since, the more we learn, the more we think it could have had a watery, perhaps life-sustaining past. The surface is scored by old gorges, canyons, beaches, ocean basins and giant volcanoes, whose eruptions could have kept things riled up on the planet. Where this water went and how, taking most of Mars's atmosphere with it, is one of the great and ominous environmental mysteries of our time.

If life did arise from those early, cozy conditions, it could have moved underground as the surface cooled and dried.

And if Mars was once flush with liquid, was it also flush with life? If astronauts ever crunch across the red sands, will they also be crunching over fossils of microbes?

The current findings, however, "cannot say anything more," Dr. Flamini said. "We may guess about what are the conditions and if the conditions are favorable."

Roberto Orosei, a co-investigator on the radar instrument and lead author of the paper published on Wednesday in the journal Science, said the scientists could not measure the thickness of the lake, but that it had to be at least a yard or so thick for the radar pulses to bounce back.

He said a back-of-the-envelope calculation indicated several hundred million cubic meters of water. That's tens of billions of gallons.

The Mars Advanced Radar for Subsurface and Ionosphere Sounding instrument, or Marsis, was developed and built by the Italians for the Mars Express mission, which entered orbit around Mars in 2003. Taking care not to damage the rest of the spacecraft, the team in charge of Marsis took two years to deploy the radar's 130-foot-long booms.

Once the instrument was working, it sent back uncertain, inconsistent findings over this polar region. But the scientists figured out how to send back the raw data to Earth. It revealed bright reflections in a triangular region as the spacecraft passed multiple times. Intense pressure of the overlying ice would warm the ice. Computer models indicate that temperatures would be about minus-90 Fahrenheit — far colder than the melting point of water. That suggests that the water is brim full of salts, allowing it to melt.

The region corresponded to a basin, adding to speculation that liquid water had flowed into this spot.

"Water tends to collect in lower topography," Dr. Orosei said.

Dr. Orosei said the scientists checked other possible explanations, like carbon dioxide ice, for the bright reflections, but those did not match the radar observations. The signals did match radar measurements of under-ice lakes in Greenland and Antarctica.

"We came thus to the conclusion that the only possible explanation for the bright reflection was the presence of liquid water," he said.

For some scientists, the bright radar reflection falls a bit short of proof.

Richard Zurek, the chief scientist in the Mars program office at NASA's Jet Propulsion Laboratory in Pasadena, Calif., said the complex, almost chaotic structure of the ice caps could affect the radar signals in unexpected ways. "You have a lot of interfaces that could do strange things to radar signals," said Dr. Zurek, who was not involved with the research. "It's the kind of signal we would expect for liquid water. Is it the only way that signal could be produced? That's the hard part."

If it is liquid water, the intense saltiness would make it hard for life, at least life as known on Earth, to survive in the lake, Dr. Lunine said. "It may exceed the salt content that any terrestrial organisms that we know of can survive in," he said.

Still, he said, "Having a stable body of liquid water today is very intriguing and worthy of study."

John Priscu, a professor of ecology at Montana State University, has been studying Antarctica biology. There, as on Mars, the surface is barren, but is more hospitable farther down. When he and his team drilled into a subsurface lake there a few years ago, they found microbes.

"They haven't seen the light of day for hundreds of thousands of years," he said. "They're eating the rocks for energy."

If it were possible to drill a mile into Mars into the newly discovered lake, he said he'd bet there was life there too. "I've been studying life in ice for 35 years," he said. "We've been finding life in places it shouldn't be according to our current thinking of life. But that's changing."

CHAPTER 4

Believers

When it comes to those who believe in aliens, individuals fall into two camps: those who believe that aliens have already made contact with humans, and those who believe we have yet to find them. In the following articles, journalists remember the lives of those who claim to have encountered extraterrestrials here on Earth, as well as some who have devoted their lives to pursuing the possibility of alien life in the universe.

Betty Hill, 85, Figure in Alien Abduction Case, Dies

BY MARGALIT FOX | OCT. 23, 2004

BETTY HILL, WHOSE ASSERTION that she was carried off by otherworldly beings in 1961 inspired a national obsession with alien abduction that remains a staple of American popular culture, died on Sunday at her home in Portsmouth, N.H. She was 85.

The cause was lung cancer, her niece Kathleen Marden said.

Mrs. Hill was not the first person to tell of an alien encounter. But her account was the first to capture the public imagination on a grand scale, defining a narrative subgenre that has flourished in the decades since.

Mrs. Hill's account was the subject of a book by John G. Fuller, "The Interrupted Journey: Two Lost Hours 'Aboard a Flying Saucer' " (Dial, 1966). In 1975 it became a television movie, "The UFO Incident."

The film starred Estelle Parsons as Mrs. Hill and James Earl Jones as her husband, Barney, who also said he was abducted.

The incident, the Hills said, occurred on the night of Sept. 19, 1961. Driving in the White Mountains of New Hampshire, they saw a light that seemed to grow larger and larger. Back home, they found what appeared to be shiny spots on the car's exterior. They could not account for a two-hour segment of their trip.

The Hills later saw a psychiatrist, who put them under hypnosis. Gradually, a narrative of the couple's lost hours emerged. They recounted many times that a group of short gray-skinned beings stopped their car and took them aboard a waiting spaceship. There, the Hills said, they were subjected to rigorous medical examinations that included inserting a long needle into Mrs. Hill's navel.

The account fit squarely in the Western narrative tradition. With a dark night, ghostly apparitions and sexual undercurrent, it had many Victorian gothic hallmarks, and it shared the common Western folklore theme of being spirited off and ravished by an otherworldly creature.

In the Hills' account, these traditional elements were transplanted to a modern but no less anxious time, the height of the cold war, when many people gazed nervously skyward.

"It's not unlike the Leda and the swan myth," said Terry Matheson, a professor of English at the University of Saskatchewan and the author of "Alien Abductions: Creating a Modern Phenomenon" (Prometheus, 1998). "The alien comes in, probes women in a distinctly sexual way for purposes that are equally inscrutable, but which may, we're told, make sense down the road."

Mrs. Hill was born Eunice Elizabeth Barrett on June 28, 1919, in Newton, N.H. A graduate of the University of New Hampshire, she was a social worker for many years. Besides her niece, survivors include three sisters, two children and three grandchildren. Mr. Hill died in 1969.

The Hills' cultural legacy includes films ("Close Encounters of the Third Kind"), television programs ("Roswell") and books, like those by Whitley Strieber and John Mack, that treat alien abduction as a plausible phenomenon.

Philip Klass, 85, Debunker of Claims of Flying Saucers, Dies

BY DOUGLAS MARTIN | AUG. 12, 2005

PHILIP J. KLASS, an electrical engineer and aviation editor who earned the nickname Sherlock Holmes of U.F.O.'s for his assiduous, acidic debunking of flying saucers and those who claim to see them, died Tuesday at his home in Merritt Island, Fla.

The cause was prostate cancer, Rosanne Klass, his daughter, said. He was 85.

Mr. Klass was for many years a top editor at Aviation Week & Space Technology magazine, where he wrote about satellite communications, lasers, missile defense systems and arms control. But he became known to a wider audience as a leading critic of the ever-growing number of Americans who are convinced that alien beings have visited Earth and taken at least a few lucky, or quite unlucky, people for a spin in space.

Mr. Klass applied careful, reasoned analysis to what is often a very emotional topic, as he repeatedly found explanations for saucer sightings in natural phenomena and manmade objects. He used details like the typefaces on documents to question the veracity of what some contended were deliberately covered-up government documents.

At a time when the cohort of people who have reported being kidnapped by aliens is growing, Mr. Klass suggested that improper use of hypnosis might be a good explanation. In his 1989 book, "UFO Abductions: A Dangerous Game," he lamented that some people "will needlessly bear mental scars for the rest of their lives" because, he said, they have succumbed to "fantasies" of alien abduction.

He wrote seven books on U.F.O.'s, appeared on many television shows and personally fielded a half-dozen phone calls a day on the subject.

His answers could be as specific as pointing to the unusual brightness of Mars and Jupiter in 1987 to explain a Japanese pilot's report of a flying saucer, and as general as his observation that people taken by aliens have never reported returning with a souvenir.

Mr. Klass could rankle people who believe in U.F.O.'s, some of whom called him "a disinformation agent," and irritate others who at least do not disbelieve. A notice about his death on the Web site of UFO Magazine referred to his "logically untenable stands on issues where even the facts were not in his favor."

Philip Julian Klass was born in Des Moines on Nov. 8, 1919, and grew up in Cedar Rapids, Iowa. He earned a degree in electrical engineering from Iowa State University and got a job with General Electric, where he worked on the development of electronics for aviation during World War II.

In 1952, Mr. Klass left G.E. to become the first avionics editor of Aviation Week & Space Technology. He retired as senior editor of the magazine in 1986 but remained a contributing editor until 2002.

For years, he offered a $10,000 prize to anybody who could provide solid scientific evidence of extraterrestrial visitations. It is still unclaimed.

In addition to his sister, who lives in Manhattan, Mr. Klass is survived by his wife, the former Nadya Gane, and a stepdaughter and stepson.

In an interview with Skeptic magazine in 1999, Mr. Klass said his "fondest hope" was that he would be abducted by aliens. Afterward, he said, he would find time to watch TV, go to movies and read books about something besides flying saucers.

When an Astronaut Believes in Aliens

THE LEDE | BY MIKE NIZZA | JULY 24, 2008

The Lede is a blog that remixes national and international news stories — adding information gleaned from the Web or gathered through original reporting — to supplement articles in The New York Times and draw readers in to the global conversation about the news taking place online.

U.F.O. SIGHTINGS ARE commonplace. Belief in aliens is high. Proof, of course, is hard to come by. In May, Jeff Peckman earned national attention for simply claiming to have a video showing an alien. The screening, though, was disappointing.

This week, another believer stepped forward to claim another round of headlines. But he wasn't shopping any proof, just a sterling background on the subject. Here's a few headlines from Google News:

— Apollo 14 Moonwalker Claims Aliens Exist
— Apollo Astronaut Edgar Mitchell Claims UFO Cover Up
— Ex-NASA scientist says aliens exist
— Moon-walker claims alien contact cover-up

Edgar Mitchell, the sixth man to walk on the moon, appeared on a British radio show to say that he was "privileged enough to be in on the fact that we've been visited on this planet and the U.F.O. phenomena is real."

A close watcher of Mr. Mitchell wouldn't have heard anything new. As Robert Pearlman, the editor of CollectSpace, wrote on a message board:

> Based on the nine minute call, it (a) doesn't seem to be anything tremendously different from prior comments and writings by Dr. Mitchell, and (b) he isn't actually claiming first-hand knowledge but rather repeating what he was told by others. It is no secret that Dr. Mitchell's Noetic Science Institute attracts those that believe in extraterrestrials and that he has attended numerous conferences where they have been on the agenda

ISAAC BREKKEN FOR THE NEW YORK TIMES

Souvenirs on display at the Little A'Le'Inn in Rachel, Nev., the heart of Alien Country.

for discussion, thus what he says here is of little surprise (and some might argue, consequence).

Indeed, Mr. Mitchell's similar remarks in the past are available on YouTube.

While Mr. Mitchell, 77, is certainly entitled to his own views on the issue, the rest of us may need to wait for something more convincing. In the end, it would seem that the latest flare-up in alien news may be explained away in exactly the same way that a Reuters headline panned a new movie based on a television show that stirred the same part of our imaginations: "Little to believe in with new 'X-Files' movie."

Budd Hopkins, Abstract Expressionist and U.F.O. Author, Dies at 80

BY MARGALIT FOX | AUG. 24, 2011

BUDD HOPKINS, a distinguished Abstract Expressionist artist who — after what he described as a chance sighting of something flat, silver, airborne and unfathomable — became the father of the alien-abduction movement, died on Aug. 21 at his home in Manhattan. He was 80.

The cause was complications of cancer, his daughter, Grace Hopkins-Lisle, said.

A painter and sculptor, Mr. Hopkins was part of the circle of New York artists that in the 1950s and '60s included Mark Rothko, Robert Motherwell and Franz Kline.

His work — which by the late '60s included Mondrian-like paintings of huge geometric forms anointed with flat planes of color — is in the collections of the Metropolitan Museum of Art and the Museum of Modern Art in New York, the Corcoran Gallery of Art in Washington and the British Museum, among others.

In later years Mr. Hopkins turned to large, quasi-architectural sculptures that seemed to spring from primordial myths. In 1985, reviewing one such piece, "Temple of Apollo With Guardian XXXXV" — it was part house of worship, part archaeological ruin, part sacrificial altar — Michael Brenson wrote in The New York Times:

> If the work is about sacrifice and violence, it is also about ecstasy and illumination. In the course of trying to re-establish the broadest meaning of the abstract geometry that has fascinated so many 20th-century artists, Hopkins makes us consider that ritual, worship, cruelty and superstition have always been inseparable.

Some articles about Mr. Hopkins made much of the relationship between these pieces and his fascination with otherworldly visitors, for by then his books, lectures and television appearances had made

CHANG W. LEE/THE NEW YORK TIMES

Budd Hopkins in Roswell, N.M., in 1997.

him well known as a U.F.O. investigator. Mr. Hopkins, however, disavowed a connection.

He was also quick to point out that he had never been abducted himself. But after what he described as his own U.F.O. sighting, on Cape Cod in 1964, he began gathering the stories of people who said they had not only seen spaceships but had also been spirited away in them on involuntary and unpleasant journeys.

As the first person to collect and publish such stories in quantity, Mr. Hopkins is widely credited with having begun the alien-abduction movement, a subgenre of U.F.O. studies. Later high-profile writers on the subject, including Whitley Strieber and the Harvard psychiatrist John Mack, credited him with having ignited their interest in the field.

In eliciting the narratives — many obtained under hypnosis — of people who said they had been abducted, Mr. Hopkins was struck by the recurrence of certain motifs: the lonely road, the dark of night, the burst of light, the sudden passage through the air and into a waiting craft, and above all the sense of time that could not be accounted for.

He went in search of that lost time. What he found, in story after story, was this:

The aliens were technically sophisticated and many spoke improbably good English. They were short, bug-eyed, thin-lipped and gray-skinned, stripped their subjects naked and probed them with instruments, often removing sperm or eggs.

These narratives, Mr. Hopkins wrote, led him to a distasteful but inescapable conclusion: The aliens — or "visitors," as he preferred to call them — were practicing a form of extraterrestrial eugenics, aiming to shore up their declining race by crossbreeding with Homo sapiens.

In 1989 Mr. Hopkins founded the Intruders Foundation, based in Manhattan, to help sound the alarm.

He wrote four books on the subject, including "Intruders: The Incredible Visitations at Copley Woods" (1987), which spent four weeks on the New York Times best-seller list and was the basis of a 1992 TV movie starring Richard Crenna.

Mr. Hopkins's work drew inevitable fire; in interviews he sometimes likened his attackers to Holocaust deniers, an analogy that incurred further criticism.

Elliott Budd Hopkins was born in Wheeling, W. Va., on June 15, 1931, and at 2 survived polio. He earned a bachelor's degree in art history from Oberlin College in 1953 and afterward settled in New York, where he soon made his artistic reputation.

After the Cape Cod sighting he described — a silvery disc over Truro, Mass. — Mr. Hopkins began researching U.F.O.'s. In 1976 he published an article about abductions in The Village Voice, which led to an article in Cosmopolitan.

The exposure drew sacks of letters from readers wondering if they too had been abducted, and his second career was born. By the 1980s, it had eclipsed the first.

Mr. Hopkins's three marriages, to Joan Baer, April Kingsley and Carol Rainey, ended in divorce. Besides his daughter, Grace, from his marriage to Ms. Kingsley, he is survived by his companion, Leslie Kean; a sister, Eleanor Whiteley; and a grandchild.

His memoir, "Art, Life and UFOs," was published in 2009 by Anomalist Books.

Unlike some writers in the genre who described their own abductions as spiritually transformative, Mr. Hopkins believed that no good could come of being the unwilling subject of a vast human genome project in the sky. He called his informants "victims" and ran group therapy sessions for them in New York.

Many who shared their stories with Mr. Hopkins had no conscious memory of their abductions at first. But they had lived for years, he said, with the nagging feeling that somewhere, something in their lives had gone horribly wrong.

Their condition, Mr. Hopkins said, was not as rare as one might suppose. By his reckoning, 1 in 50 Americans has been abducted by an alien and simply does not know it.

John Billingham, Seeker of Extraterrestrials, Dies at 83

BY WILLIAM YARDLEY | AUG. 10, 2013

DR. JOHN BILLINGHAM, who as a NASA official in the 1970s helped persuade the federal government to use radio telescopes to scour the universe for evidence of extraterrestrial intelligence, even as critics mocked the idea, died on Aug. 4 in Grass Valley, Calif. He was 83.

His death was confirmed by his sons, Robert and Graham.

Dr. Billingham, an Englishman who earned a medical degree at Oxford and helped design spacesuits for astronauts in the 1960s, never found the evidence he was looking for. But he did help establish the validity of the quest.

"We sail into the future, just as Columbus did on this day 500 years ago," Dr. Billingham said on Oct. 12, 1992, when after two decades of planning and maneuvering NASA formally began its search for extraterrestrial intelligence, known by the acronym SETI. "We accept the challenge of searching for a new world."

The effort, which Dr. Billingham led as chief of the life sciences division at NASA's Ames Research Center in California, involved using huge radio telescopes to search for radio signals — either deliberate intergalactic flares or incidental noise — emitted by other technologically advanced civilizations that might be billions of years old and billions of light-years away.

"The whole picture is that we are the newcomers on the block, that they're out there, other civilizations that are much older than we are," Frank Drake, a radio astronomer who in 1960 started seeking signals from beyond the solar system, said in an interview. "Anybody we find would probably be way ahead of us in longevity and probably in sophistication."

Yet a year after NASA began the project, SETI lost its federal financing amid Congressional assertions that it was a waste of taxpayer money — "a great Martian chase" in the words of one critic, Senator Richard H. Bryan, a Nevada Democrat.

Dr. Billingham retired not long after, but neither he nor SETI was finished.

Operating as the nonprofit SETI Institute, based in Mountain View, Calif., Dr. Billingham and a team of scientists cobbled together financing from universities and high-tech billionaires to keep the effort going. The Allen Telescopic Array, jointly owned by the institute and the University of California, Berkeley, is named for Paul G. Allen, a co-founder of Microsoft, who gave $25 million to the cause.

Although the federal government no longer pays SETI scientists to search for intergalactic radio signals, federal grants have helped pay for some of the SETI equipment used in recent years. Government emphasis has shifted toward another endeavor Dr. Billingham supported, which is also pursued by scientists at the institute: the rapidly expanding field of astrobiology, which includes searching for extraterrestrial life at the most microbial level, not just forms that might transmit radio signals.

Dr. Billingham first learned of astrobiology, then called exobiology, in 1968, through the work of the astronomer and author Carl Sagan and others.

"It changed my whole life," he once wrote.

Three years later, he recruited Barney Oliver, the research chief of Hewlett-Packard, to host a symposium at which they and others sketched out a plan for using a $10 billion array of giant radio telescopes to search for extraterrestrials. They called it Project Cyclops.

"We are almost certainly not the first intelligent species to undertake the search," they wrote in a proposal that spanned more than 200 pages. "The first races to do so undoubtedly followed their listening phase with long transmission epochs, and so have later races to enter the search. Their perseverance will be our greatest asset in our beginning listening phase."

Dr. Billingham was born on March 18, 1930, in Worcester, England. He completed his medical studies at Oxford in 1954 and later spent six years as a medical officer in the Royal Air Force. He joined NASA in 1963, becoming chief of its environmental physiology branch later that year at the Johnson Space Center in Houston. He moved to the Ames Research Center in 1965 and spent the next several years in NASA's biotechnology divisions while he built support for SETI.

In addition to his sons, he is survived by four grandchildren. His wife, the former Margaret Macpherson, also a physician, died in 2009.

SETI was not formally incorporated into Dr. Billingham's official job title at NASA until March 1991, when he became chief of the space agency's Office of the Search for Extraterrestrial Intelligence. When financing was eliminated three years later, he became a senior scientist at the SETI Institute.

One of Dr. Billingham's concerns was how to respond to a radio signal from space. To answer the question, he helped draft the "Declaration of Principles Concerning Activities Following the Detection of Extraterrestrial Intelligence." The document allowed that a proper response would depend on the signal received. Only so much advance planning is possible.

"A lot of people think this is silly, but we need to give a lot of thought to a reply," Dr. Billingham said in 1992. "It is not a question just for scientists and engineers. Already we agree on one rule: Don't reply unless you have undertaken extensive international consultation."

Ionel Talpazan, Whose U.F.O. Art Had Sightings All Over, Dies at 60

BY WILLIAM GRIMES | SEPT. 29, 2015

IONEL TALPAZAN, an outsider artist from Romania who sold his visionary works of U.F.O.s and life in outer space on the sidewalks of Manhattan before being discovered in the late 1980s, died on Sept. 21 in Manhattan. He was 60.

The cause was complications of a stroke and advanced diabetes, Aarne Anton, his dealer at the American Primitive Gallery in SoHo, said.

Mr. Talpazan claimed that one night in the Romanian countryside, when he was 8, a strange, hovering shape slowly descended from the sky, enveloping him in a celestial blue light, and then disappeared. The experience haunted him and became the source of his art.

His paintings, drawings and sculptures dealt, obsessively, with U.F.O.s and their inner workings, often shown in cross section and heavily annotated in Romanian.

He insisted that his work had value not only for art lovers but also for NASA scientists, since it articulated the magnetic forces and antimatter at work in the propulsion systems of his spaceships.

"My art is about the big mystery in life," he told the journal Western Folklore in 2008. "How did we get here on planet Earth? Why are we here? Is there life on other planets?"

Ionel Talpazan (pronounced yah-NEL TAL-puh-zan) was born on Aug. 16, 1955, in the commune of Petrachioaia, Romania. After being given up by his parents, he was raised by foster parents in Maineasca, one of the commune's four villages.

His close encounter with a U.F.O. occurred when, fearing a beating for misbehavior, he slipped out of his bedroom window in the middle of the night and walked out into the surrounding countryside, where he stood transfixed by what he called "a blue energy" radiating from a mysterious source overhead.

The incident left him confused, but also deeply interested in the idea of space travel, and he set about rendering his interplanetary visions, especially U.F.O.s, on paper. "I felt that by drawing them, I might penetrate their mystery," he told The Independent of London in 1996.

He escaped from Romania, where he had worked in the construction trade, by swimming across the Danube to Yugoslavia in 1987. After several months living in a United Nations camp in Belgrade, he was granted political asylum by the United States and emigrated to New York.

A television documentary on U.F.O.s rekindled his interest in space, and he began drawing hypothetical interplanetary spacecraft. "He was interested not so much in aliens as in otherworldly technology," said Daniel Wojcik, whose book "Outsider Art, Trauma and Visionary Worlds" will be published by the University Press of Mississippi next year. "He thought flying saucers would help bring about a better world by introducing a benevolent technology."

In New York, Mr. Talpazan lived hand-to-mouth, at times sleeping in a cardboard box near Columbus Circle. He sold his work on the sidewalk, becoming a familiar sight at the entrance to the annual Outsider Art Fair, then held in the Puck Building in SoHo.

He was discovered by the art dealer Henry Tobler, known as Jay, who saw him selling work outside the Museum of Modern Art and wrote about him in 1990 in Folk Art Messenger, the journal of the Folk Art Society of America.

In the 1996 exhibition "Visions of Space & UFOs in Art," at the American Primitive Gallery in Manhattan, more than a dozen of Mr. Talpazan's works covered one wall, some of them bought by the artists Brice Marden and Terry Winters. The following year, at the same gallery, he was the subject of a solo show, "Ionel Talpazan: U.F.O.: Art & Science," and at the Musée d'Art Brut in Neuilly-sur-Marne, France.

Mr. Talpazan rendered his U.F.O.s in various guises. Some adhered to an illustrational realism; others were abstract and heavily patterned like mandalas. Some works showed a single U.F.O. lifting off from an unidentified planet. Others showed multiple saucers engaged in battle or disappearing into a wormhole.

His titles were matter-of-fact yet otherworldly: "Red UFOs and the Statue of Liberty," "Father and Son in Space," "UFOs Over NYC."

His U.F.O. sculptures, a little wider than a Frisbee, were made of plaster and painted silver or blue, then outfitted with brightly colored portholes and exhaust pipes and set on pedestals made from scavenged parts.

In the New Jersey newspaper The Star-Ledger in 1997, the critic Dan Bischoff wrote that they resembled "those old metal tops that you sent spinning by pushing a spiral rod down into the center, big and rounded and Art Deco-looking, like two hubcaps from a Studebaker stuck together."

Mr. Talpazan's work was exhibited at the American Visionary Art Museum in Baltimore, the Yerba Buena Center for the Arts in San Francisco and the Martin-Gropius-Bau in Berlin.

In 2013, work by Mr. Talpazan was included in "The Alternative Guide to the Universe" at the Hayward Gallery in London and "Farfetched: Mad Science, Fringe Architecture and Visionary Engineering" at the Gregg Museum of Art and Design in Raleigh, N.C. This year he was part of the exhibition "Arstronomy" at La Casa Encendida in Madrid.

Mr. Talpazan, who lived in Harlem, is survived by two brothers and two sisters. About a year ago, on taking American citizenship, he legally changed his name to Adrian DaVinci.

"My art shows spiritual technology, something beautiful and beyond human imagination, that comes from another galaxy," he told Western Folklore. "Something superior in intelligence and technology. So, in relative way, this is like the God. It is perfect."

The Woman Who Might Find Us Another Earth

BY CHRIS JONES | DEC. 7, 2016

The star-crossed life of Sara Seager, an astrophysicist obsessed with discovering distant worlds.

LIKE MANY ASTROPHYSICISTS, Sara Seager sometimes has a problem with her perception of scale. Knowing that there are hundreds of billions of galaxies, and that each might contain hundreds of billions of stars, can make the lives of astrophysicists and even those closest to them seem insignificant. Their work can also, paradoxically, bolster their sense of themselves. Believing that you alone might answer the question "Are we alone?" requires considerable ego. Astrophysicists are forever toggling between feelings of bigness and smallness, of hubris and humility, depending on whether they're looking out or within.

One perfect blue-sky fall day, Seager boarded a train in Concord, Mass., on her way to her office at M.I.T. and realized she didn't have her phone. She couldn't seem to decide whether this was or wasn't a big deal. Not having her phone would make the day tricky in some ways, because her sons, 13-year-old Max and 11-year-old Alex, had a soccer game after school, and she would need to coordinate a ride to watch them. She also wanted to be able to find and sit with her best friend, Melissa, who sometimes takes the same train to work. "She's my best friend, but I know she has other best friends," Seager said, wanting to make the nature of their relationship clear. She is an admirer of clarity. She also likes absolutes, wide-open spaces and time to think, but not too much time to think. She took out her laptop to see if she could email Melissa. The train's Wi-Fi was down. She would have to occupy herself on the commute alone.

Seager's office is on the 17th floor of M.I.T.'s Green Building, the tallest building in Cambridge, its roof dotted with meteorological and radar equipment. She is a tenured professor of physics and of planetary science, certified a "genius" by the MacArthur Foundation in 2013. Her area of expertise is the relatively new field of exoplanets: planets that orbit stars other than our sun. More particular, she wants to find an Earthlike exoplanet — a rocky planet of reasonable mass that orbits its star within a temperate "Goldilocks zone" that is not too hot or too cold, which would allow water to remain liquid — and determine that there is life on it. That is as simple as her math gets.

Her office is spare. There is a set of bookshelves — "Optics" and "Asteroids III" and "How to Build a Habitable Planet" — topped with a row of certificates and honors leaning against a chalkboard covered with equations. In addition to the MacArthur award, which doesn't come with a certificate but with $625,000, she is proudest of her election to the National Academy of Sciences. Although the line between lunacy and scientific fact is constantly shifting, the search for aliens still occupies the shadows of cranks, and Seager hears from them almost daily, or at least her assistant does. By the standards of her universe, Seager is famous. She is careful about the company she keeps and the words she chooses. She isn't searching for aliens. She's searching for exoplanets that show signs of life. She's searching for a familiar blue dot in the sky.

That means Seager, who is 45, has given herself a very difficult problem to solve, the problem that has always plagued astronomy, which, at its essence, is the study of light: Light wages war with itself. Light pollutes. Light blinds.

Seager has a commanding view of downtown Boston from her office window. She can sweep her eyes, hazel and intense, all the way from the gold Capitol dome to Fenway Park. When Seager works at night and the Red Sox are in town, she sometimes has to close her curtains, because the ballpark's white lights are so glaring. And on this morning, after the sun completed its rise, her enviable vista became unbearable. It was searing, and she had to draw her curtains. That's how light can be the object of her passion and also her enemy. Little lights — exoplanets — are washed out by bigger lights — their stars — the way stars are washed out by our biggest light, the sun. Seager's challenge is that she has dedicated her life to the search for the smallest lights.

THE VASTNESS OF SPACE almost defies conventional measures of distance. Driving the speed limit to Alpha Centauri, the nearest star grouping to the sun, would take 50 million years or so; our fastest current spacecraft would make the trip in a relatively brisk 73,000 years. The next-nearest star is six light-years away. To rocket across our galaxy would take about 23,000 times as long as a trip to Alpha Centauri, or 1.7 billion years, and the Milky Way is just one of hundreds of billions of galaxies. The Hubble Space Telescope once searched a tiny fragment of the night sky, the size of a penny held at arm's length, that was long thought by astronomers to be dark. It contained 3,000 previously unseen points of light. Not 3,000 new stars — 3,000 new galaxies. And in all those galaxies, orbiting around some large percentage of each of their virtually countless stars: planets. Planets like Neptune, planets like Mercury, planets like Earth.

As late as the 1990s, exoplanets remained a largely theoretical construct. Logic dictated that they must be out there, but proof of their existence remained as out of reach as they were. Some scientists dismissed efforts to find exoplanets as "stamp collecting," a derogatory term within the community for hunting new, unreachable lights just to name them. (Even among astronomers, there can be too much stargazing.) It wasn't until 1995 that the colossal 51 Pegasi b, the first widely recognized exoplanet orbiting a sunlike star, was found by a pair of Swiss astronomers using a light-analyzing spectrograph. The Swiss didn't see 51 Pegasi b; no one has. By using a complex mathematical method called radial velocity, they witnessed the planet's gravitational effect on its star and deduced that it must be there.

There has been an explosion of knowledge in the relatively short time since, in part because of Seager's pioneering theoretical work in using light to study the composition of alien atmospheres. When starlight passes through a planet's atmosphere, certain potentially life-betraying gases, like oxygen, will block particular wavelengths of light. It's a way of seeing something by looking for what's not there.

Light or its absence is also the root of something called the transit technique, a newer, more efficient way than radial velocity of finding exoplanets by looking at their stars. It treats light almost like music, something that can be sensed more accurately than it can be seen. The Kepler space telescope, launched in 2009 and now trailing 75 million miles behind Earth, detects exoplanets when they orbit between their stars and the telescope's mirrors, making tiny but measurable partial eclipses. A planet the size of Jupiter passing in front of its sun might result in a 1 percent dip in the amount of starlight Kepler receives, a drop that, in time, reveals itself to be as regular as rhythm, as an orbit. The transit technique has led to a bonanza of finds. In May, NASA announced the validation of 1,284 exoplanets, by far the largest single collection of new worlds yet. There are now 3,414 confirmed exoplanets and an additional 4,696 suspected ones, the count forever increasing.

Before Kepler, the nature of the transit technique meant that most of those exoplanets were "Hot Jupiters," giant balls of hydrogen and helium with short orbits, making them scalding, lifeless behemoths. But in April 2014, Kepler found its first Earth-size exoplanet in its star's habitable zone: Kepler-186f. It's about 10 percent larger than Earth and orbits on the outer reaches of where the temperature could allow life. No one knows the mass, composition or density of Kepler-186f, but its discovery remains a revelation. Kepler was searching, somewhat blindly, an impossibly small sliver of space, and it found a potentially habitable world more quickly than anyone might have guessed.

In August, astronomers at the European Southern Observatory announced that they had detected a somewhat similar planet orbiting Proxima Centauri, the single star closest to us after the sun. They named it Proxima Centauri b. Studying the data, Seager supported the discovery and agreed that it might boast a life-sustaining — or at least non-life-threatening — surface temperature. There are now nearly 300 confirmed exoplanets or candidates orbiting within the habitable zones of their stars. Extrapolating the math, NASA scientists now believe that there are tens of billions of potentially life-sustaining planets in the Milky Way alone. The odds practically guarantee that a habitable planet is somewhere out there and that someone or something else is, too.

In some ways, the search for life is now where the search for exoplanets was 20 years ago: Common sense suggests a presence that we can't confirm. Seager understands that we won't know they're out there until we more truly lay eyes on their home and see something that reminds us of ours. Maybe it's the color blue; maybe it's clouds; maybe, however many generations from now, it's the orange electrical grids of alien cities, the black rectangles of their lightless Central Parks. But how could we ever begin to look that far? "Everything brave has to start somewhere," Seager says.

The beginning of her next potential breakthrough hangs on the wall opposite the window in her office. It is a two-thirds scale model of a single petal of something called the starshade. She has been a leading proponent of the starshade project, and outside her teaching, it is one of her principal professional concerns.

Imagine that far-off aliens with our present technology were trying to find us. At best, they would see Jupiter. We would be lost in the sun's glare. The same is true for our trying to see them. The starshade is a way to block the light from our theoretical twin's sun, an idea floated in 1962 by Lyman Spitzer, who also laid the groundwork for space telescopes like Hubble. The starshade is a huge shield, about a hundred feet across. For practical reasons that have to do with the bending of light, but also lend it a certain cosmic beauty, the starshade is shaped exactly like a sunflower. By Seager's hopeful reckoning, one day the starshade will be rocketed into space and unfurled, working in tandem with a new space telescope like the Wfirst, scheduled to launch in the mid-2020s. When the telescope is aimed at a particular planetary system, lasers will help align the starshade, floating more than 18,000 miles away, between the telescope and the distant star, closing the curtains on it. With the big light extinguished, the little lights, including a potential Earthlike planet and everything it might represent, will become clear. We will see them.

The trouble is that sometimes the simplest ideas are the most complicated to execute. About once a decade since Spitzer's proposal — he could work out the math but not the mechanics — someone else has taken up the cause, advancing the starshade slightly closer to reality before technological or political inertia set in. Three years ago, Seager joined a new, NASA-sponsored study to try to overcome the final practical hurdles; NASA then chose her from among her fellow committee members to lead the effort.

After those decades of false starts, Seager and her team have already succeeded in making the starshade seem like a real possibility. NASA recognized it as a "technology project," which is astral-bureaucracy speak for "this might actually happen." Today the starshade is a piece of buildable, functional hardware. Seager packs that single petal into a battered black case and wheels it, along with a miniature model of the starshade, into classrooms and conferences and the halls of Congress, trying to find the momentum and hundreds of millions of dollars that allow impossible things to exist.

"If I want the starshade to succeed, I have to help mastermind it," Seager says. "The world sees me as the one who will find another Earth." She has her intelligence, and her credentials, and her audience. She has her focus. But maybe more than anything else, Seager understands in ways few of us do that sometimes you need darkness to see.

SEAGER GREW UP in Toronto, wired in a way all her own. "Ever since I was a child, there was just something about me that wasn't quite like the others," she says. "Kids know how to sort through who's the same and who's different." After her parents divorced, her father, Dr. David Seager, achieved a certain fame by becoming one of the world's leaders in hair transplants. The Seager Hair Transplant Center still operates and bears his name a decade after his death. David Seager was besotted with his bright daughter and wanted her to become a physician.

Seager did her best to fit in. Sometimes she did; mostly she didn't. Eventually, she gave up trying. She still talks breathlessly — "without enough modulation," she has learned by listening to other people talk. She has never had the patience to invest in something like watching TV. "Things just move too slowly," she says. "It feels like a drag." She sleeps a lot, but that's just a concession to her biology; she recognizes that she's a more efficient machine when she's rested. But if Seager's apartness didn't make her insecure, it also made her feel as though the expectations of others didn't apply to her. "I loved the stars," she says. When she was 16, she bought a telescope.

Friendless for most of her childhood, Seager eventually forged her way to her own vision of the good life. She found and married a quiet man named Mike Wevrick, whom she met on a ski trip with her canoe club. He had seen something in her that nobody other than her father fully saw; he saw her as special as well as strange. Later, she graduated from Harvard, an early expert in exoplanets. (51 Pegasi b was discovered just when she was searching for a thesis topic. "I was born at the perfect time," she says.) She and Wevrick had Max and Alex; Seager was hired by M.I.T., and she and Wevrick and the boys moved into a pretty yellow Victorian in Concord, Mass. She took the train to work. Wevrick, a freelance editor, managed just about everything that didn't involve the search for intelligent life in the universe. Seager never shopped for groceries or cooked or pumped gas. All she had to do was find another Earth.

Then, in the fall of 2009, Wevrick got a stomachache that drove him to bed. They figured it was the flu. Wevrick didn't have the flu, but a rare cancer of the small intestine. They were told that the initial prospects were good, and he fought the cancer sufferer's systematic fight. But while laws govern astrophysics, cancer is an anarchist. About a year after Wevrick's diagnosis, he and Seager went cross-country skiing, and he couldn't keep up. A few more terrible months passed, and he began writing out a methodical three-page list, practical advice for Seager after his death. It wasn't a love letter; it was an instruction manual for life on Earth. By June 2011, he was 47 and in home hospice. Seager asked him how to get the roof rack that carried his canoes off the car. "It's too complicated to explain," Wevrick said. That July, he died.

The first couple of months after Wevrick's death were weird. Seager felt a surprising sense of relief from the uncertainties of sickness, a kind of liberation. She didn't care about conventions like money, which she had never needed to manage, and she took the boys on some epic trips. There are pictures of them smiling together in the deserts of New Mexico, on mountaintops in Hawaii. Then one day, she went into Boston for a haircut, got turned around and accidentally walked into a lawyer's office next to the salon. Seager ended up talking to a woman inside. That woman was also a widow, and she told Seager that there would be a moment, as inevitable as death itself, when her feelings of release would be replaced by the more lasting aimlessness of the lost. Seager walked back outside, and just like that, the world came out from under her feet. She fell into an impossible blackness.

Later that winter, she took the boys sledding at the big hill in Concord. Two other women and their children were there. Seager stared at them coldly. They were smiling and carefree with their perfect, blissful lives. Seager felt ugly and ruined next to them. Then Alex, who was 6 at the time, had a meltdown. He sprawled himself across the hill so that the other children couldn't go down it. The two other mothers tried to get him to move. "He has a problem," Seager told them. They continued to try to shift him.

"HE HAS A PROBLEM," Seager said. "MY HUSBAND DIED."

"Mine, too," one of the other women said. That was Melissa. A few weeks later, on Valentine's Day, Seager was invited to her first gathering of the widows.

Today, Melissa says she could detect the telltale "flintiness" of the recently bereaved the moment she saw Seager on the hill. Now there were six widows united in Concord, each middle-aged, each in a different stage of grief, drawn together by the peculiar pull of the unlucky. Three had been widowed by cancer, two by accidents — bicycling and hiking — and one by suicide. Melissa's husband was four years gone, Seager's seven months.

Widowhood was like a new universe for Seager to explore. She had never understood many social norms. The celebration of birthdays, for instance. "I just don't see the point," she says. "Why would I want to celebrate my birthday? Why on earth would I even care?" She had also drawn a hard line against Christmas and its myths. "I never wanted my kids to believe in Santa." After Wevrick's death, she became even more of a satellite, developing a deeper intolerance for life's ordinary concerns.

Making dinner seemed an insurmountable chore, the routine of school lunches a form of torture. The roof needed to be replaced, and she didn't have the faintest idea how to get it fixed. She wasn't sure how to swipe credit cards. If the answers to her questions weren't somewhere on Wevrick's three wrinkled sheets of paper, it could feel as though they were locked in a safe.

There was a pendant light in her front hall, where the boys would fight with their toy lightsabers, and sometimes they would hit the light with their wild swings. Seager decided that either the light or one of the boys was going to end up damaged. She asked the widows how to do electrical work — "I have to parcel out things with logic and evidence," she says — got out the ladder and took down the light, carefully wrapping black tape around the ends of the bare wires that now poked through the hole in the ceiling. She remembers thinking that her removing that light, all by herself, represented the height of her new accomplishment. She felt so reduced. She felt so gigantic.

FOR ALL OF HER real and perceived strangeness, the most unusual thing about Seager is her blindness to her greatest gift. She is more than aware of her preternatural mathematical abilities, her possession of a rare mind that can see numbers and their functions as clearly as the rest of us see colors and shapes. "I'm good at that stuff," she says with her brand of factual certainty that is sometimes confused with arrogance. She knows she is unusually capable of turning abstract concepts into things that can be packed into a case. What she doesn't always see is her knack for connection between places if not always people, the unconventional grace she possesses when it comes to closing unfathomable distances.

Seager has lined the hallway outside her office with a series of magical travel posters put out by the Jet Propulsion Laboratory. Each gives a glimpse of the alien worlds that, in part because of her, we now know exist. There's a poster for Kepler-16b, an exoplanet that orbits a pair of stars, like Luke Skywalker's home planet of Tatooine. Kepler-186f is depicted with red grass and red leaves on its trees, because its star is cooler and redder than the sun, which might influence photosynthesis in foliage-altering ways. There's even one for PSO J318.5-22, a rogue planet that doesn't orbit a star but instead wanders across the galaxy, cast in perpetual darkness, swept by rain of molten iron.

After the discovery of Proxima Centauri b, Seager wrote a galactic postcard from it for the website Quartz. She closed her eyes and imagined a world 25 trillion miles away. "For the average earthling," she wrote, "visiting this planet might not be much fun." She saw a planet perhaps a third larger than Earth, with an orbit of only 11 days. Given its proximity to its small, red star, she suggested that the ultraviolet radiation on Proxima Centauri b is probably intense but the light Martian-dim. She also deduced that Proxima Centauri b is "tidally locked." Like the moon's relationship to Earth, one side of the planet always faces its star, which is always in the same place in its sky. Parts of Proxima Centauri b are cast in perpetual sunrise or sunset. One side is always in darkness.

At first, after Wevrick's death, Seager thought about abandoning her work, because she was having such a hard time with her responsibilities at home. Her dean talked her out of quitting, giving her financial support to hire caregivers for the boys and urging her to redouble her efforts. "I had worked so hard," she says. "I had all the years I called the lost years with Mike when I ignored him. We had little tiny kids. I was working all the time, exhausted all the time. But I was like: We'll have money some day. We'll have time some day."

She paused. Her face was blank, emotionless. "Now I'll cry." Seconds later, tears spilled out of her eyes, and her voice modulated. "I wanted to make it up to him, and I never did."

Seager has always found comfort and perhaps even solace in her work, in her search for another and maybe better version of our world. In her mourning, each discovery represented one more avenue of escape. In the spring of 2013, she was given responsibility for the starshade. That July, she met a tall, fast-walking man named Charles Darrow.

Darrow, who is now 53, was an amateur astronomer and the president of the Toronto branch of the Royal Astronomical Society of Canada, and at the last minute he decided to go to the society's annual meeting in Thunder Bay, Ontario. Darrow was on his way out of a profoundly unhappy marriage; he worked for his family business, an engine-parts wholesaler. He needed a break, and he pointed his car north. "I wanted to be alone," he says. At a reception on the Friday evening, Darrow noticed a hazel-eyed woman staring at him from across the room. "I thought she was looking at someone behind me," he says. Then he went into the lecture hall, and the same woman was that night's keynote speaker. She talked about exoplanets. The next day, lunch was in a university cafeteria. The woman was in the salad line ahead of him, and she turned around. Darrow mustered up his courage and invited Sara Seager to join him. "I knew about five minutes into the conversation that my life was going to change," he says.

Seager was taken with Darrow the night she saw him in Thunder Bay. She had been struck by the contrast between the whiteness of his shirt and his tanned summer skin. But she didn't have the same certainty that possessed him at their lunch the next day. She wasn't sure how to develop a relationship across the 549 miles between her home in Concord and his home outside Toronto. She thought they might never cross paths again.

They might not have, except Darrow resolved during his drive back home that he had to call her. He picked up the phone five times but always hung up before she answered. On the sixth, he spoke to her, beginning a long correspondence, emails and conversations over Skype. Darrow and Seager talked every way but face to face. They fell in love remotely. "I had to follow my heart," Darrow says. "I decided that I wasn't going to die unhappy."

Melissa, meanwhile, told Seager that if she could close the gap between here and a planet like Kepler-186f — a journey that would take us 500 light-years to complete — then the 549 miles between Concord and Toronto shouldn't seem like such an insurmountable gulf. By her usual measures, he was right next door.

Seager and Darrow married in April 2015. In different ways, each had rescued the other. Seager was the cataclysm that allowed Darrow to make every correction. He divorced, left his family business and moved into a pretty yellow Victorian in Concord. The two boys started calling him dad. For Seager, Darrow was a second chance to know love, even deeper than the one she had known, because it seemed so improbable in her sadness. "I feel so lucky to have found him," Seager says. "What are the chances?"

Adapting to his new life hasn't always been easy for Darrow. He is determined, as he puts it, "to make Sara the happiest woman in the multiverse." He cooks dinner; he helps take care of the boys; he maintains the house; he walks with Seager to the train station every morning, and he picks her up every night. He has chosen to take care of the mundane so that she can devote herself to the extraordinary. But he banged his head more than once on Wevrick's canoe, which still hung from the back of the garage.

Not long ago, Darrow was looking for the right ways to assert his presence, to make a claim to a house that didn't always feel like his. The wires dangling from the front hall ceiling bothered him. They looked bad and seemed dangerous. A few months after his arrival in Concord, he took his opening. He carved out some of the plaster, installed a plastic box, ran the wires through it and hooked up a new fixture, flush mounted, so that the boys wouldn't hit it during their duels.

Darrow climbed down from the ladder and flicked the switch.

THE MORNING AFTER she forgot her phone, Seager woke up and decided, just like that, to skip the commute. With the house to herself, she tried to make coffee. She left out part of the machine, and after some terrible noises, the pot was bone dry. She sat down at her kitchen table with her empty mug and began talking about hundreds of billions of galaxies and their hundreds of billions of stars. Tens of billions of habitable planets, far more of them than there are people on Earth. There has to be other life somewhere out there. We can't be that special.

"It would be arrogant to think so," Seager said. But in her lifetime, after the Wfirst telescope rockets into orbit, and maybe her starshade follows it — she puts the chances of success at 85 percent — she will have time to explore only the nearest hundred stars or so. A hundred stars out of all those lights in the sky, a fraction of a fraction of a fraction.

Will one of them have a small, rocky planet like Earth? Probably. Will one of those small, rocky planets have liquid water on it? Possibly. Will the planet sustain life? Now the odds tilt. Now they are working against her, and she knows it. Now they're maybe one in a million that she'll find what she's looking for.

She did some private math. "I believe," she said.

Seager's discovery will be fate-altering if it comes, but it will also be quiet, a few pixels on a screen. It will obey the laws of physics. It will be a probability equation: What are the chances? We won't discover that there is life on other planets the way we've been taught that we'll learn. There won't be some great mother ship descending from the sky over Johannesburg or a bizarre lightning storm that monsters will ride to New Jersey. What Seager will have is a photograph from a space telescope of a distant solar system, with its star eclipsed by her starshade, and with a familiar blue dot some safe and survivable distance away from it. That's all the evidence she will have that we're not alone, and that will be all the evidence she will need. Her proof of life will be a small light where there wasn't one before.

CHRIS JONES is a two-time National Magazine Award winner for feature writing.

Winston Churchill Wrote of Alien Life in a Lost Essay

BY KIMIKO DE FREYTAS-TAMURA | FEB. 15, 2017

LONDON — Even as he was preparing for the biggest struggle of his life, leading Britain in its fight against Nazi Germany, Winston Churchill had something else on his mind: extraterrestrials.

In a newly unearthed essay sent to his publisher on Oct. 16, 1939 — just weeks after Britain entered World War II and Churchill became part of the wartime cabinet — and later revised, he was pondering the likelihood of life on other planets.

Churchill, who went on to become prime minister during much of World War II and again from 1951 to 1955, was so enthralled by the subject that he even ordered a suspected sighting of an unidentified flying object by the Royal Air Force to be kept a secret for 50 years to avoid "mass panic."

In an 11-page essay titled "Are We Alone in the Universe?" the statesman showed powers of reason "like a scientist," said Mario Livio, an astrophysicist who read the rarely seen draft and wrote about it in an article published on Wednesday in Nature magazine.

"The most amazing thing is that he started this essay when Europe was on the brink of war and there he is, musing about a question about a scientific topic that is really a question out of curiosity," he said in an interview.

Churchill first defines what life is, then details the requirements for life to exist and progressively expands his reasoning to the existence of life in other solar systems, Mr. Livio said. "He's really thinking about this," Mr. Livio said, "and though he didn't have all the knowledge at hand, he thinks about this with the logic of a scientist."

Churchill's interest in science stemmed from his early years as an army officer in British-ruled India, where he had crates of books, including Darwin's "On the Origin of Species," shipped to him by his mother.

He later became friends, at least for a time, with the writer H.G. Wells, whose novel "The War of the Worlds," about Martians invading Britain, had been adapted by Orson Welles for a famous CBS radio broadcast in 1938 — a year before Churchill wrote his article. (Churchill once said Wells's "The Time Machine" was one of the books he would like to take with him to Purgatory.)

Churchill argued that it was probable that extraterrestrial life existed somewhere in the universe. This was years before Frank Drake, the American astronomer and astrophysicist, presented in 1961 his theory about the number of communicative civilizations in the cosmos. "It is astonishing that Churchill wasn't a scientist and yet he showed such an interest in science," Mr. Livio said.

The manuscript was passed on to the National Churchill Museum in Fulton, Mo., the site of Churchill's famed 1946 Iron Curtain speech, in the 1980s by Wendy Reves, the wife of Churchill's publisher, Emery Reves. It had been overlooked for years until Timothy Riley, who became the museum's director last year, stumbled upon it recently. Soon after news of the discovery, two other copies were found in a separate archive in Britain.

Although the article was sent to Mr. Reves in 1939, it was not published. Churchill had revised it a number of times in the 1950s.

In his article, Churchill wrote: "I am not sufficiently conceited to think that my sun is the only one with a family of planets."

"I, for one, am not so immensely impressed by the success we are making of our civilization here that I am prepared to think we are the only spot in this immense universe which contains living, thinking creatures," he wrote, "or that we are the highest type of mental and physical development which has ever appeared in the vast compass of space and time."

Largely self-educated in the sciences, Churchill had boundless curiosity for practically anything, an attitude he once described as "picking up a few things as I went along."

He wrote about 30 million words in his lifetime, including wartime speeches, an African travelogue, a book on oil painting, a lengthy memoir, and even an essay on an imagined invasion of Russia when he was just 15. For his body of work, he won the Nobel Prize for Literature in 1953.

Welding an active imagination with scientific thought, Churchill produced a few madcap ideas — which he called "funnies" — that he actually championed while he was prime minister, as a means to defeat Nazi Germany.

There was Operation Habakkuk, an imagined fleet of aircraft carriers made from wood pulp and ice to fight German U-boats in the mid-Atlantic. Then there was the Great Panjandrum, an enormous, rocket-propelled wheel packed with explosives. Churchill even invented a green velvet "siren suit" to be put on in a hurry during air raids.

While none of these ideas came into being (the giant wheel having run amok in the testing stage), science was not just a hobby for Churchill. He was the first prime minister to hire a science adviser. Frederick Lindemann, a physicist, became Churchill's "on tap" expert and once described him as a "scientist who had missed his vocation," said Andrew Nahum, who organized an exhibition on Churchill and science at the Science Museum in London. He found a separate copy of the essay in the Churchill Archives Center at the University of Cambridge.

Churchill also met regularly with scientists such as Bernard Lovell, the father of radio astronomy and the Lovell telescope.

"Churchill presided over a culture that encouraged technological development," Mr. Nahum said. Churchill had such a genuine interest in science, he added, that as chancellor of the Exchequer in prewar Britain, he complained to a friend of having to draft the budget instead of reading a book on quantum physics.

During World War I, when he was lord of the admiralty and later secretary of state for air and war, he encouraged military aviation, chemical warfare and tanks. During World War II, which he called in his memoirs "The Wizard War," he supported the development of radar, rockets and Britain's nuclear program.

Churchill founded in 1958 the British equivalent of the Massachusetts Institute of Technology at Cambridge — Churchill College — which has since produced 32 Nobel Prize winners.

In the interwar period, Churchill wrote numerous scientific articles, including one called "Death Rays" and another titled "Are There Men on the Moon?" In 1924, he published a text asking readers "Shall We All Commit Suicide?" in which he speculated that technological advances could lead to the creation of a small bomb that was powerful enough to destroy an entire town.

Churchill's recently unearthed article on extraterrestrial life was probably written in the same vein and was probably intended to be published as a popular science piece for a newspaper.

Two other scientific essays — one on cell division in the body and another on evolution — are stored in the museum's archives in Fulton, Mr. Riley, the museum director, said in an interview.

Churchill had a "natural curiosity and general optimism about life," Mr. Riley said. He had "a willingness to see technical and scientific advances improve not only his immediate world or his country, but the world."

Ursula Marvin, Geologist of the Extraterrestrial, Dies at 96

BY RICHARD SANDOMIR | MARCH 9, 2018

URSULA MARVIN WAS a geologist at the Smithsonian Astrophysical Observatory in Cambridge, Mass., when she and her colleagues were asked to examine an extraterrestrial object: a 10-pound chunk of Sputnik IV, a Soviet satellite that had crashed, superheated at 1,535 degrees Celsius, onto a street in Manitowoc, Wis., before dawn on Sept. 5, 1962.

Investigating a fragment from Sputnik IV — a less heralded part of the space program that had begun with the thunderclap of the first Sputnik's orbits of Earth in 1957 — proved irresistible to a mineralogical expert.

Clinging to one end of the fragment, Dr. Marvin found, were droplets of minerals, including wustite, an iron oxide.

"Until then, wustite had been viewed as an artificial product too unstable to survive in nature," she said in an oral history interview in 2013 with Derek Sears of NASA's Ames Research Center. But when she X-rayed samples of several meteorites in the Harvard Museum, she found wustite in all of them, confirming that they were formed through iron oxidation during re-entry into Earth's atmosphere.

"The wustite had always been there," she said, "but nobody had X-rayed meteorite fusion crusts before."

Dr. Marvin — who would later hunt for meteorites in Antarctica and analyze moon rocks from Apollo missions — died on Feb. 12 at a nursing home in Concord, Mass., a niece, Gayl Bailey Heinz said. She was 96.

Ursula Alice Bailey was born in rural Bradford, Vt., on Aug. 20, 1921. Her father, Harold, was a government entomologist (one newspaper called him the state's "official bug hunter"), and her mother, the former Alice Bartlett, was a schoolteacher.

Young Ursula's love of the outdoors was sparked while growing up near the White Mountains of New Hampshire, where sunsets "shone with a pink-purple afterglow," she said in a lecture in 1997 to the Harvard-Smithsonian Center for Astrophysics, a joint venture of the Smithsonian observatory and the Harvard College Observatory.

But she did not became interested in geology until she was a student at Tufts University, where she majored in history but was required to study science for two years. Biology bored her; geology transformed her.

"Here was a professor talking about mountains, how they form and change, about rivers, lakes, deserts, beaches, dunes and how the earth itself formed and evolved," she said in the lecture. "I never knew there was such a science."

Inspired, she asked her geology professor if she could change her major.

He rebuffed her, telling her that she should be learning to cook.

Undaunted, she added geology, math and physics courses. The experience of being one of the few female geologists at the time led her, decades later, to advance the cause of female scientists, in part as chairwoman of the women's program committee at the astrophysics center.

After she graduated, she earned a master's degree in geology from Radcliffe College. Following World War II, she moved to Chicago to be a research associate at the University of Chicago while her husband, Lloyd Chaisson, attended dental school.

But their marriage was short-lived, and she returned to Harvard to study for her Ph.D. While there, she met another geology student, Thomas Marvin. They married, and before she had finished her doctorate, Mr. Marvin was hired by a company to prospect for ore deposits in Brazil and Angola. The expeditions, which they undertook together starting in 1953, lasted several years.

Recalling their work in Corumbá, in southwestern Brazil, Dr. Marvin said their job was to search for manganese to use in making steel. "When the water was up, we traveled across country by dugout canoe (sometimes decked out in wild orchids)," she said in the lecture.

They returned full time to the United States in 1958. After teaching mineralogy at Tufts for two years, she was offered a job researching meteorites at Harvard before joining the Smithsonian observatory in 1961.

At the time, Dr. Marvin knew nothing about meteorites. But with the space age accelerating, she gladly accepted. She received her doctorate from Harvard in 1969.

As her meteoritic expertise grew exponentially in the late 1970s, Dr. Marvin joined the first of her three expeditions to Antarctica to hunt for meteorites. She was the first woman on the American research teams that traveled there.

"To search for Antarctic meteorites is an exhilarating adventure," she wrote in New Scientist magazine in 1983. She described riding a snowmobile over blue ice and drifting snow: "The glimpse of a dark object starts the heart pounding. Racing toward it, the excitement grows as one sees it is not a shadow, not a glacial cobble, but a meteorite — a piece of rock from another planet."

In 1982, she was part of the team that discovered a lunar meteorite.

"Many miles from camp," she wrote in New Scientist, they found a small specimen with a "frothy, greenish-tan crust" that was unlike anything they had ever seen. When it was later analyzed in Washington, it was found to bear a great similarity to rocks that had been found by astronauts in the lunar highlands.

Following her trips to Antarctica, a small mountain on the ice sheet was named for her (Marvin Nunatak). And her meteorite studies earned her a similar honor in 1991, when the International Astronomical Union named an asteroid for her.

Dr. Marvin's interest in meteorites knew no geographic boundaries. When a six-pound meteorite ripped through the roof of a house in Wethersfield, Conn., in 1982, she and other scientists arrived to inspect it the next day. It was the second meteorite strike in Wethersfield in 11 years. "Meteorites are always a dramatic occurrence," Dr. Marvin told The New York Times, "but to have two strike the same town is, well, almost incomprehensible."

She said the second meteorite, which had rolled to a stop under Wanda and Robert Donahue's dining room table, was probably from an asteroid belt between Mars and Jupiter that she called a "sort of celestial rock garden."

More critically, she was part of a study, funded by the National Aeronautics and Space Administration, that analyzed rocks and soil recovered by Apollo astronauts and Soviet robotic missions.

"We compared rock types from different sites on the moon to work out its geological history," she said in the oral history. Her boss at the astrophysics center, John Wood, "concluded the moon must have had an early magma ocean," she said, adding, "I think that idea still is the best explanation of how the lunar crust was formed."

She described her findings in an article for Science magazine in 1989. In all, she wrote more than 160 research papers and a book, "Continental Drift: The Evolution of a Concept" (1973).

Dr. Wood, a former associate director of the astrophysics center, said in a telephone interview: "We were a small group — just four of us and several technical assistants — that was very passionate about these samples. They were heady times, and Ursula was the mineralogical arm of the team."

She leaves no immediate survivors. Her husband died in 2012.

After the Apollo 17 mission in 1972 — the crew included Harrison Schmitt, a geologist — Dr. Marvin was in Houston when a box of rocks the astronauts had collected was opened. It was the final Apollo mission.

"There was so much of interest to study in the lunar samples," she said in 1997, "that I continued to do it until year before last."

CHAPTER 5

Encounters

Reports of alien sightings are many and varied. From the stories of individuals who report being abducted by aliens to obsessions with crop circles, humans find ways of seeing aliens everywhere. In these articles, journalists follow possible reportings of alien encounters, from formations that inspire alien comparisons to a mummy found in Chile that seemed to have alien DNA. Whether or not any of these constitute evidence of extraterrestrial life remains to be seen.

Despite Lack of Data From Pilots and Officials, Reports of UFO Sightings Are Many and Widespread

BY WALTER SULLIVAN | OCT. 21, 1973

RARELY, IF EVER, since Kenneth Arnold reported in 1947 seeing what came to be known as "flying saucers" during flight near Mount Rainier in Washingon State have there been such widespread reports of unidentified flying objects, or UFO's, as in recent days.

They ranged from Rochester, N.Y., where a flying V formation of lights was reported, to Gulfport, Miss., where a press account told of "strange creatures with weirdly shaped heads" stopping cars on Route 90 "and scratching at the windows." Two men in Pascagoula, Miss., even said they had been taken aboard a UFO by creatures with crab-claw hands.

Relatively well documented was evidence that two objects, possibly aircraft or meteorites, flew at supersonic speed across the Northern United States. Tremors characteristic of a sonic boom were recorded by earthquake detectors at Pennsylvania State University in State College, Pa., at 8:53 P.M. on Oct. 11 and 1:26 P.M. on Oct. 17.

One or both of these booms were recorded by similar instruments at the State University of New York in Binghamton, at Virginia Polytechnic Institute in Blacksburg and on an air pressure recorder in Michigan.

LIGHT EVADES COPTER

In the vicinity of Dover, Del., three women last Sunday night told the state police they had been watching a brilliant light in the sky for 45 minutes. A police helicopter was sent to investigate and according to a spokesman at Dover Air Force Base, the police and a man in the air base control tower saw the light, too, but the helicopter was unable to overtake it or determine its distance.

In Louisiana, sheriff's deputies reported chasing five orange-red lights for 12 miles through the piney woods. An Indiana man said a nighttime UFO "followed me home," and elsewhere policemen said they had been "buzzed" by a swooping UFO while on patrol.

Yet the Federal Aviation Administration, which is responsible for air traffic control over the United States, said at the week's end that its radar network had seen nothing unusual. Its 90 long-range radars and 130 airport radars cover about 90 per cent of American air space above 24,000 feet and lesser amounts at lower levels.

NO REPORTS BY PILOTS

The F.A.A. added that no reports of special significance had been submitted by airline pilots, although it was noted that in recent years pilots had tended to refrain from making such reports because so much paperwork was involved.

An airline source said that pilots in recent years have reported UFO's at a rate of about two a month. These reports are channeled to the Center for Short-Lived Phenomena of the Smithsonian Astrophysical Observatory in Cambridge, Mass. However, it was reported that there had been no marked increase in such reports during recent weeks.

A spokesman for the North American Air Defense Command denied that any UFO had been detected in the last three weeks. The command, known as NORAD, operates a space tracking system that monitors earth satellites and watches for incoming missiles.

A number of last week's reports originated in Ohio near Wright-Patterson Air Force Base, which, for many years, was the headquarters for Air Force cataloguing of UFO reports. In 1966 the Air Force asked the University of Colorado to conduct an independent investigation of such reports to assess their significance.

SCOFFS AT 'VISITORS'

The two-year study was directed by Dr. Edward U. Condon, an internationally known physicist and former head of the National Bureau of Standards. Dr. Condon's salty comments soon antagonized those inclined to take seriously the possibility that UFO's are visitations from other worlds.

"If you define a UFO as a visitor from outer space," he recently told United Press International, "there's no evidence they exist. I've never seen one. I think further study of UFO's would be scientifically useless. I think my own study of UFO's was a waste of Government money."

A contrary view has been maintained by Dr. J. Allen Hynek, head of the Dearborn Observatory of Northwestern University. Dr. Hynek was consultant to the Air Force UFO project, which was terminated after the Condon report had downgraded the value of such investigations.

Dr. Hynek and a few others have fought a relatively lonely battle for renewed efforts to investigate the possibility that some phenomenon of significance lies behind the many and varied reports.

LIBRARY OF CONGRESS PRINTS AND PHOTOGRAPHS DIVISION

The northeast view of Wright-Patterson Air Force Base, Area B, Building 51 in Dayton, Ohio.

One explanation for many UFO reports is the "chasing" phenomenon experienced by motorists, particularly when bright planets are visible on clear nights, such as those of recent days. Four planets are exceptionally bright at this time:

Venus is in the western sky, shortly after sunset. Jupiter is higher in the evening sky. Mars, to the east in the early evening is reddish and exceptionally bright, having made a close approach to the earth this week. Saturn, which rises later, is also exceptionally near the earth.

The "chasing" effect, which has induced panic in many drivers since "flying saucers" first were publicized, occurs when one sees a planet or very bright star from a moving car. Every time the car turns, speeds up or slows down, the planet appears to do the same, in contrast to nearby landscape features that move past the observer in the expected fashion.

The planets, at their brightest, change color as their light is modified by atmospheric conditions, particularly near the horizon. Such color changes are often a feature of UFO's.

PRANKS ARE GENERATED

A flurry of UFO reports not only breeds more reports but also gives birth to a variety of pranks. In Shreveport, La., where 5,000 people gathered Wednesday night for a UFO "fly-in," according to The Associated Press, they were rewarded by the startling sight of a large red object passing overhead. It proved to be a balloon released as a joke.

Responding to a rash of UFO reports in Indiana, the police intercepted three plastic garbage bags that had been made into glowing hot-air balloons by suspending candles beneath them.

A similar hoax, while the investigation was being conducted at the University of Colorado in Boulder, sent flickering UFO's across the town. However, it led to a police warning that such pranks could be incendiary.

In Greenwood, Del., United Press International reported, a traffic jam developed as drivers stared at a saucer-shaped circle of lights that proved to have been erected by volunteer firemen, using their emergency generator as a power source. Five of them were charged with disorderly conduct.

SYMBOLIC LIGHTS

Another circle of 92 flashing lights was set up in Texas to entice a UFO within photographic range. A single powerful light emitted three short flashes and one long one and the "whole system" was designed to symbolize the hydrogen atom, indicating a knowledge of science by the inhabitants of earth.

Preparations to receive a UFO were also made at Palacios, a small town on the Texas coast. The Mayor, W. C. Jackson, said, according to United Press International, "It just occurred to me that no one has ever made those fellas welcome." Hence the town council issued a welcoming proclamation.

Two Texans might claim having seen the most clearly identified UFO to date. It carried red and white flashing lights, they said, and the inscription "U.F.O." on one side.

While the Condon study attributed conventional explanations to most sightings, there was a small residue that remained perplexing. Some involved seemingly reliable observers and could not be dismissed out of hand.

MORE DATA NEEDED

The gist of the findings was, however, that such episodes did not indicate visitations from afar and could have been explained in conventional ways had more been known about them.

One oft-cited series of sightings, which demonstrated the fallibility of human observation, occurred on March 3, 1968, in the region between Kentucky and Pennsylvania.

An Indiana woman told the Air Force: "All of the observers saw a long jet-airplane-looking vehicle without any wings." It had many windows, she said, and was at treetop level, "seen very clearly and was just few yards away."

In Ohio a woman told how her dog had crawled between garbage cans in the driveway as the UFO passed overhead, whimpering "like he was frightened to death." She herself, she added, felt an "overpowering drive to sleep."

Elsewhere in Ohio a business machine company executive told how a UFO had chased him down the highway, making every turn that he did and staying directly over his car as he sought to escape.

From Tennessee the Air Force received the report of a woman who said she could see rivets on the metallic fuselage of the UFO. She was terrified because she believed it to be close by, yet it flew in complete silence. The most accurate reports came from airline pilots who recognized that the phenomenon was far above them.

All of the sightings were traced to a single event. Earlier on the day of their occurrence Moscow had announced the launching of Zond IV, a spacecraft destined to be fired into the "outlying regions of near-earth space" from a parking orbit around the earth.

The launching failed, Zond IV plunged back into the atmosphere over the United States and broke into a procession of fiery fragments that constituted the "windows" seen by many observers, none of whom was probably less than 100 miles from the re-entry trajectory.

A U.F.O., or Is There an Explanation?

BY ALBERT J. PARISI | JAN. 1, 1995

ON OCT. 30, ROBERT F. COAR SAYS, the lights he saw in the night sky were not stars — and the experience has haunted his dreams and waking moments ever since. What he saw, he says, was an unidentified flying object that passed over his Fifth Avenue neighborhood in Paterson, N.J., pursued by helicopters.

"I couldn't believe what I was seeing, but there it was, this thing about the size of a tractor-trailer, shaped like a shoe box with pulsating red and orange lights," said Mr. Coar, 35, a self-employed sign painter and mural artist. "It was about 100 feet up, about level with an old smokestack back there."

Mr. Coar reported his sighting to a neighbor, Debbie Bodensky, an office manager for a cemetery, and she sent him packing. "How could I believe something so loony as that?" she said.

Ten minutes later, driving with a relative to a local shop, "we saw it, exactly as Bob described," she said. "It had these big orange and yellow lights, and it didn't make a sound."

Mr. Coar's account is not unique. George A. Filer of Medford, the state director of the Mutual U.F.O. Network, an international research group based in Seguin, Tex., said that since 1990 New Jersey has averaged at least 12 U.F.O. sightings a year that warranted the group's investigation. Mr. Coar's sighting is one of them.

"The reports vary and many times involve multiple witnesses," said Mr. Filer, a sales manager with a local utilities auditing company and a former United States Air Force intelligence officer. "Many sightings can be explained away relatively easily. What could be an aircraft, a hoax or a natural phenomenon such as a light reflection or even an unusual cloud formation — this sort of thing we have answers for."

The group's 150 New Jersey researchers discount up to 90 percent of reported sightings, Mr. Filer said, adding: "It's that 10 percent that is the most intriguing for us. That's where our detective work and resources come together."

For each U.F.O. sighting called to the group's attention, he said, 100 go unreported because "no one wants to be labeled as a nut."

In the case of Mr. Coar's sighting, Paterson police officials acknowledged receiving dozens of phone calls that evening reporting odd activities in the sky, but they say this had nothing to do with U.F.O.'s. On that night, two small helicopters dispatched from the state National Guard worked in conjunction with police on an anti-drug mission.

The flight, which lasted nearly five hours, resulted in about 60 drug-related arrests on Oct. 30. Reports of U.F.O. sightings that night, Police Chief Richard Munsey said, are the result of "overactive imaginations."

New Jersey ranks 14th in the nation in overall U.F.O. investigations with some 38 between 1987 and 1993. Indiana and Florida are tied for first place with 283 investigations each in the same period. According to the Mutual U.F.O. Network, the higher numbers reflect expanded reporting networks in both those states. In the last four years, New Jersey members have received some 225 reports of sightings.

Mr. Filer is the host of a weekly two-hour program on WWAC entitled "Investigations — U.F.O.," broadcast on a UHF channel from Atlantic City.

One of the most dramatic sightings occurred in South Bound Brook last March 16 when a woman, driving with her two grown daughters, stopped at a traffic light on Queens Bridge above the Raritan River. They reported seeing a U.F.O. with flashing multicolored lights hovering at window level above the river.

It was so close, Mr. Filer said, "they could have hit it with a rock." The family was "terrorized," he said, and reported the incident to the police. Later, he said, the family learned from a friend who was listening to a police scanner that a report was broadcast advising officers to "keep an eye out for flying saucers" and that anyone spotting it "should give it a speeding ticket."

The police explanation of his experience has not satisfied Mr. Coar. "I know what I saw, and there was more to it than what the police are saying," he said.

Ms. Bodensky agrees. She said she and her companion had driven to within 100 feet of the object and observed it for several minutes. "We just knew that it wasn't a helicopter," she said.

Mr. Coar says the incident has changed his life in ways he could never have predicted. The frustration of getting people to believe him, for example, has continued. "I still hope others will come around and talk about what they saw, what I saw," he said.

And he cited another long-lasting effect: "For me, there has always been something appealing about the night, something mystifying about the stars. But after that night, the stars, to me at least, will have a quality of fear."

Recent sighting reports and field investigations carried out by New Jersey members of the research organization have ranged from mundane to "downright bizarre," said Peter A. Jordan of Clark, the director for Union County. Mr. Jordan, who holds a master's degree in psychology from the New School for Social Research in Manhattan and a Bachelor of Arts degree in philosophy from Drew University in Madison, lectures on the paranormal at Fordham, Drew and other universities in the area.

Mr. Jordan says some U.F.O. activity can be tied into paranormal phenomena traditionally labeled as hauntings or poltergeist activity. His research has led him to believe that people who report sightings and U.F.O. abductions are often telepathic, have a history of repressed memory or have suffered child abuse or sexual abuse in their formative years.

"All too often, we have seen examples of this in our investigations," Mr. Jordan said. "We can't explain why, not entirely, but it is more than a random pattern."

U.F.O. Hits Congestion at O'Hare, Turns Back

BY TOM ZELLER JR. | JAN. 2, 2007

A COLUMN IN The Chicago Tribune yesterday has had many skywatchers atwitter over the last 24 hours, in that it details the apparent sighting by several United Airlines workers at Chicago's O'Hare airport of ... something hovering over Concourse C on Nov. 7.

The official word from both the F.A.A. and from United executives, of course, is that there's nothing to see here folks, move along, but Jon Hilkevitch, who writes the paper's "Getting Around" column, interviews several employees who said otherwise:

A United manager said he ran outside his office in Concourse B after hearing the report about the sighting on an internal airline radio frequency.

"I stood outside in the gate area not knowing what to think, just trying to figure out what it was," he said. "I knew no one would make a false call like that. But if somebody was bouncing a weather balloon or something else over O'Hare, we had to stop it because it was in very close proximity to our flight operations."

The money quotes, however, appeared to come from a "taxi mechanic" who described something that would raise the hairs on anyone's neck:

"I tend to be scientific by nature, and I don't understand why aliens would hover over a busy airport," said a United mechanic who was in the cockpit of a Boeing 777 that he was taxiing to a maintenance hangar when he observed the metallic-looking object above Gate C17.

"But I know that what I saw and what a lot of other people saw stood out very clearly, and it definitely was not an [Earth] aircraft," the mechanic said.

For what it's worth, this appears to be the same taxi mechanic who filed a report with something called The National U.F.O. Reporting Center on Nov. 21 — about two weeks after the incident allegedly occurred.

An excerpt from the mechanic's account:

I work for a major airline at O'Hare, I am a taxi mechanic. I have the job responsibility of moving aircraft under their own power from gate to gate or the hangar complex for maintenance. We also accomplish the engine run-up testing needed. So I hope that does something for establishing a little of credibility for my report. I am still in absolute wonder and amazement at what I saw that afternoon.

Around 16:30 a pilot made a comment on the radio about a circle or disc-shaped object hovering over gate C-17 at the C concourse in Chicago. At first we laughed to each other and then the same pilot said again on the radio that is was about 700-feet a.g.l. (above ground level). The day was overcast with the ceiling being reported at 1600 feet if I remember correctly. I was taxiing a Boeing 777 from the Int'l Terminal to the Company Hangar on the North side of the Airport. As we passed the C Terminal on the Alpha taxi-way we observed a dark gray hazy round object hovering over O'Hare Int'l Airport. It was definitely over the C Terminal. It was holding very steady and appeared to be trying to stay close to the cloud cover. The radio erupted with chatter about the object and the A.T.C. controller that was handling ground traffic made a few smart comments about the alleged UFO sighting above the C terminal.

We had to continue moving the aircraft to the hangar. After parking I noticed the craft [was] no longer there but there was an almost perfect circle in the cloud layer where the craft had been. The hole disappeared a few minutes later.

For the rest of the night there were jokes made on the radio about the sighting.

We don't quite know what to make of any of this, but a spokeswoman for the Federal Aviation Administration, Elizabeth Isham Cory, apparently offered Mr. Hilkevitch of The Tribune this view: blame the weather.

"Our theory on this is that it was a weather phenomenon," Ms. Cory said. "That night was a perfect atmospheric condition in terms of low [cloud] ceiling and a lot of airport lights. When the lights shine up into the clouds, sometimes you can see funny things. That's our take on it."

Formations in China Desert Are Still a Mystery

THE LEDE | BY J. DAVID GOODMAN | NOV. 18, 2011

The Lede is a blog that remixes national and international news stories — adding information gleaned from the Web or gathered through original reporting — to supplement articles in The New York Times and draw readers in to the global conversation about the news taking place online.

IN THE AREAS of the Internet that buzz with delight at strange formations writ large on the landscape, a mystery appeared to be solved this week: giant shapes seen on Google Maps in China are simply geometric targets for satellite calibration.

Or are they?

The shapes, which include thick white lines drawn in sharp angles and structures arranged in concentric circles around airplanes, attracted attention earlier this month after they were first highlighted by the online magazine, Viewzone. That post includes a photo of one such formation, partially built, from 2003 and what appears to be a village nearby.

It was later picked up by Gizmodo, the technology blog, which posted several other images found by readers. The blog observed that the forms, some up to a mile long, "seem to be designed to be seen from orbit."

The massive forms in the Gobi Desert in western China appear to be several years old and previously attracted the attention of at least one blogger, who in March 2009 wrote: "There's something screwy going on in western China."

As far back as 2004, they seem to have drawn more than just casual interest, as Wired magazine's Danger Room blog recently reported in a post after the formations came to light again this week. The post draws on data gathered by enabling a feature of Google Earth that shows the date and size of the satellite images stitched together to create a seamless image by the service. The fact that many images of the remote location were taken over time, beginning in 2004, suggests a costly interest on the part of an individual or organization.

As former CIA analyst Allen Thomson notes, turning on the DigitalGlobe coverage layer in Google Earth shows all the various times the imaging satellite has been asked to inspect that part of the desert. "Starting in 2004, somebody has ordered many, many satellite pictures of it," Thomson tells Danger Room. "Can't have been cheap."

On Twitter-like social media sites in China, users shared links to articles in the Western press (China's official media have not yet weighed in) and pet theories about what in the world they could be.

"If you like crop circles, Area 51 or UFOs in the United States, then you will love this," one user of the popular Chinese microblog Sina Weibo wrote.

"In fact, as early as 2008, China's major media reported on the shooting range in the northwestern desert state," another stated, without elaborating, in a comment along with a reposted CNN video.

Amid the theories both in and out of China, the mystery seemed to find a mundane explanation on Thursday.

A site called Life's Little Mysteries said the forms were "almost definitely used to calibrate China's spy satellites," citing comments in an interview with a research technician at the Mars Space Flight Facility at Arizona State University.

The satellite calibration theory was advanced by a blogger in 2008 to explain crop circles, which have also been the subject of fevered online speculation.

But The Lede found that after consulting with independent and United States government satellite experts, the calibration theory may be incorrect.

"With calibration, you're looking for precise measurement," said Dwayne Day, a military space historian, in a telephone interview. "You have boxes that get smaller by a calculated amount. You don't just throw stuff all over the place and then take a picture of it."

He said that when calibration targets have been used by the United States and Russia, they are much smaller. "There's no reason why you would build anything that big for a satellite calibration target," he said.

Decades-old markings that are more likely candidates for satellite targets can still be seen on Google Maps in Arizona. A video of one of these markings, which resemble cross hairs, was posted by an amateur historian who said they were likely part of recently declassified American spy satellite programs from the 1960s and early 1970s. But that site too may have nothing to do with space or satellites, analysts said; they could be related to military aviation.

After an email inquiry by The Lede, the Union of Concerned Scientists, an industry watchdog and critic, said the China formations appeared to be conventional aerial and missile bombing targets. In the past, China has built large structures for bombing practice.

In short, the China sites remain a mystery — but not necessarily one with an out-of-this-world explanation.

"The thing that would make it really sexy is if there were fences around it — and I don't see any," Mr. Day, the historian, said, adding that a lack of security indicated a lack of strategic importance. "We don't know what the heck it is, but there are probably two guys in China who could tell you what it is and you'd be bored silly."

Bright Lights, Strange Shapes and Talk of U.F.O.'s

BY JONAH ENGEL BROMWICH | NOV. 12, 2015

WHEN STRANGE THINGS appear in the sky, many people can't help but turn their thoughts to extraterrestrials. But there's usually a more down-to-earth explanation.

That was the case when a bright light in the sky off the Southern California coast last weekend touched off a flurry of excitement about unidentified flying objects.

After news reports, the Navy reluctantly confirmed it had been testing a Trident II (D5) missile fired from a submarine. A second and final missile was tested on Monday, The Los Angeles Times reported.

It was one of several recent sightings in the sky to cause talk about U.F.O.s. Others included a group of strangely shaped clouds over Cape Town, also over the weekend, and an Army veteran's claim that he spotted a "solid, dark-gray triangle-shaped craft" in the sky over Portland, Tenn., last week. Most of these sightings go unreported in the mainstream news media, though a variety of blogs and sites track them.

"The mind abhors a vacuum of explanation," said Michael Shermer, 61, the publisher of Skeptic magazine and a columnist for Scientific American. "Short of a good explanation, people just turn to the one that most immediately comes to mind which, in pop culture, is extraterrestrials."

A poll of a random sample of 1,114 American adults conducted by National Geographic in 2012 found that 77 percent believed "there are signs that aliens have visited Earth." (It also found that President Obama would "handle an alien invasion" better than Mitt Romney, who was running for president at the time.)

Another, more rigorous survey from Time and CNN conducted in 2000 found that 20 percent of respondents said they knew someone who had seen a U.F.O.

Peter Davenport, 67, is the director of a two-person organization called the National U.F.O. Reporting Center in Washington State. He compiles reports on sightings, like one in 2013 that came from a former astronaut, Byron Lichtenberg, who is based in North Texas.

(Texas is a hotbed of U.F.O. sightings, including many that turn out to be the atmospheric phenomenon known as the Marfa Lights.)

Mr. Lichtenberg confirmed that he did call in the report, though he pointed out by email that a few months after his sighting, details began to emerge about the Lockheed Martin SR-72 aircraft.

"It would make sense if that's what we saw," he said.

Mr. Davenport, of the reporting center, was dismissive of the idea that the naval exercise near Los Angeles should even be discussed in the same breath with possible alien sightings.

"We're struggling with a semantic issue here," he told me. "The term U.F.O. from my vantage point alludes to a genuine alien craft that has exhibited flight characteristics that are altogether incompatible with terrestrial aircraft or any kind of object of terrestrial origin."

Mr. Davenport sent links to several reports of the kind of phenomena he was interested in, and described in detail the Phoenix Lights incident, playing a recording from a witness over the phone.

Fife Symington, the former governor of Arizona, initially denied having seen the mysterious lights that floated over Phoenix in 1997. He eventually confirmed to various news organizations that he had seen them, calling them "otherworldly."

Mr. Davenport said that his principal responsibility was to avoid misleading people with data that had "nothing to do with the authentic U.F.O. cases." But he said that many of the reports he receives are "from sincere and qualified witnesses that are seeing something that is dramatically bizarre."

The scene Mr. Davenport describes in Phoenix is an echo of several famous films about U.F.O.'s, like "Close Encounters of the Third Kind," Steven Spielberg's 1977 contribution to alien mythology. The cultural landscape has been saturated with stories about U.F.O.'s since the beginning of the Cold War and the "first stirrings of the space age," as Mr. Shermer, the Skeptic publisher put it.

The popular "X Files" series, due for a six-episode return in January, popularized the idea that the government was hiding secrets about alien technology. Many episodes of the show opened with the tagline, "The Truth Is Out There."

More likely, the truth is in our heads. Mr. Shermer, who is also the author of a book called "The Believing Brain," is of the opinion that the possibility of alien life speaks to a spiritual need, calling it "almost a replacement for mainstream religion."

"In a way, extraterrestrials are like deities for atheists," he said. "They're always described as these vastly superior, almost omnipotent beings coming down from on high, very much like the Christ story, the Mormon story or the Scientology story."

Although he is about as professional a skeptic as it is possible to be, Mr. Shermer said that he remained interested in the "supernatural, the paranormal, science and religion, God, extraterrestrials, U.F.O.'s, ESP."

He added, "It is all fascinating and, if it were true, it'd be fantastic."

People Are Seeing U.F.O.s Everywhere, and This Book Proves It

BY RALPH BLUMENTHAL | APRIL 24, 2017

SYRACUSE, N.Y. — Why have sightings of unidentified flying objects around the nation more than tripled since 2001? Why is July the busiest month for U.F.O. sightings? Why did they spike in Texas in 2008, or in New Mexico in September 2015?

And how in the world, or out of it, has Manhattan racked up New York State's second-highest tally of U.F.O. sightings in this century?

These questions and many others emerge from the first comprehensive statistical summary of so-called close encounters: 121,036 eyewitness accounts, organized county by county in each state and the District of Columbia, from 2001 to 2015.

The unlikely compendium, "U.F.O. Sightings Desk Reference," is the work of a couple in Syracuse, who crunched unruly data on U.F.O. reports collected by two volunteer organizations: the Mutual U.F.O. Network, or Mufon, and the National U.F.O. Reporting Center, or Nuforc.

It is the reference "U.F.O. researchers dreamed of having," Gordon G. Spear, emeritus professor of physics and astronomy at Sonoma State University in California, writes in the foreword.

The book contains no narrative or anecdotal accounts, just 371 pages of charts and graphs that slice and dice the geography and timing of the incidents and the various shapes that witnesses reported: flying circles, spheres, triangles, discs, ovals, cigars.

Many of the sightings turn out to be explainable, the authors say, but a small percentage defy resolution.

The authors are Cheryl Costa, 65, a former military technician and aerospace analyst, and her wife, Linda Miller Costa, 62, a librarian at Le Moyne College and a former librarian at the National Academy of Sciences, NASA and the Environmental Protection Agency.

HEATHER AINSWORTH FOR THE NEW YORK TIMES

Cheryl Costa, left, and Linda Miller Costa, the authors of "U.F.O. Sightings Desk Reference."

Working on PCs amid sewing tables in the upstairs parlor — the warmest room in their hundred-year-old house — the two spent weekends for the last 16 months extrapolating figures from sightings reports and laying out the graphics.

Cheryl Costa was writing New York Skies, a U.F.O. blog for The Syracuse New Times, when the Costas decided to expand their tallies of U.F.O. sightings nationwide. "We wanted to do our bit for disclosure," she said. "It's something the government should have been doing."

The Costas realize some might find this a strange way to spend weekends. But both say they have spotted U.F.O.s themselves and want to detoxify the subject.

"We're doing scientific research," Cheryl Costa said. "What's crazy is not being willing to look at research."

She came to the collaboration roundabout, having served as a cable lineman in the Air Force in Vietnam, and afterward in the Navy's submarine service, as a man before undergoing gender-reassignment surgery in the 1980s. Ordained as a Buddhist nun, she was running a theater group in Maryland when she met Linda. They wed in 2011.

U.F.O. trackers welcomed their publication.

"With this compendium, Cheryl and Linda Costa have reminded the public and the media the extraterrestrial phenomenon continues unabated," said Stephen Bassett, founder and executive director of the Paradigm Research Group, which lobbies for disclosure of official U.F.O. records.

Rebutting a common perception that U.F.O. sightings are on the wane, the Costas' book shows that sightings have risen in waves, to 11,868 nationwide in 2015 from 3,479 in 2001. Only a small fraction of sightings are actually reported to Mufon or Nuforc.

Their labor of love is about the numbers, just the numbers, and the Costas refrain from speculating on what exactly is happening. "We really don't know," Linda Costa said. "But all these people are seeing these things."

The government officially quit the U.F.O. business in 1968, with the finding in the Condon report from the University of Colorado that there was nothing significant to investigate, although some 30 percent of the incidents were unexplained.

Mufon's 500 volunteer investigators, however, continue to check out many of the sightings reported to the group. Roger Marsh, a Mufon spokesman, said that of the 270 cases his group investigated in Manhattan from 2002 through 2016, 44 eluded explanation and remained "unknown."

One of the most intriguing occurred on the afternoon of Sept. 17, 2011, when a man on the roof terrace of the New Museum on the Bowery photographed a fast-moving diamond-shaped object with windows and flashing blue and red lights against the TriBeCa skyline.

According to Mufon, it resembled an unknown flying object photographed in Round Rock, Tex., two weeks earlier.

The Costas listed 426 sightings in New York County from 2001 to

2015, second in the state's tallies only to Suffolk County, on the tip of Long Island, with 554. How so many sightings in the nation's densest core and around its toniest beach resorts have escaped wider notoriety is just part of the mystery.

For the U.F.O. enthusiast, the pages of graphs and charts are a treasure trove of hard-to-find detail.

The District of Columbia, with 9,856 people per square mile, had the fewest sightings: 154. (A political snub from deep space?) Wyoming, with 5.8 people per square mile, had more than twice as many: 337.

Fireballs made up nearly 8 percent of the sightings in Indiana (230) and fewer than 5 percent in Colorado (157).

California, the most populous state, led the nation in U.F.O. reports (15,836, more than the next two states, Florida and Texas, combined). Los Angeles County alone had more sightings than 40 states, followed by Maricopa County, Ariz., which includes Phoenix.

Population fails to explain the figures conclusively, the Costas said. Washington State, with 6.7 million people according to the 2010 census, ranks No. 4 in sightings, ahead of Pennsylvania, with 12.7 million people, and New York State, with 19 million.

Rather, the Costas theorize, the figures may reflect good West Coast weather, which draws more people outside where they may spot U.F.O.s. Nationwide sightings peak in July, they found, and drop off between December and February.

Still, in Mississippi, U.F.O. reports spike in January and November; in New Mexico, in September.

The arduous breakdown by the nation's more than 3,000 counties was notable for revealing clusters of sightings in remote regions, places where U.F.O.s are almost never mentioned. But every county in the United States appears to have seen at least one U.F.O.

In the end, the Costas noted, the spikes may have a lot to do with media coverage.

Was a Tiny Mummy in the Atacama an Alien? No, but the Real Story Is Almost as Strange

BY CARL ZIMMER | MARCH 22, 2018

NEARLY TWO DECADES AGO, the rumors began: In the Atacama Desert of northern Chile, someone had discovered a tiny mummified alien.

An amateur collector exploring a ghost town was said to have come across a white cloth in a leather pouch. Unwrapping it, he found a six-inch-long skeleton.

Despite its size, the skeleton was remarkably complete. It even had hardened teeth. And yet there were striking anomalies: it had 10 ribs instead of the usual 12, giant eye sockets and a long skull that ended in a point.

Ata, as the remains came to be known, ended up in a private collection, but the rumors continued, fueled in part by a U.F.O. documentary in 2013 that featured the skeleton. On Thursday, a team of scientists presented a very different explanation for Ata — one without aliens, but intriguing in its own way.

Ata's bones contain DNA that not only shows she was human, but that she belonged to the local population. What's more, the researchers identified in her DNA a group of mutations in genes related to bone development.

Some of these mutations might be responsible for the skeleton's bizarre form, causing a hereditary disorder never before documented in humans.

Antonio Salas Ellacuriaga, a geneticist at the University of Santiago de Compostela in Spain who was not involved in the new study, called it "a very beautiful example of how genomics can help to disentangle an anthropological and archaeological dilemma."

"DNA autopsies," as Dr. Ellacuriaga calls them, could help shed light on medical disorders "by looking to the past to understand the present."

The research, published in the journal Genome Research, began in 2012, when Garry P. Nolan, an immunologist at Stanford University, got wind of the U.F.O. documentary, "Sirius," while it was still in production.

Dr. Nolan emailed the producers and offered to look for DNA in the mummy. The skeleton's owner agreed to X-ray images as well as bone marrow samples taken from the ribs and right humerus.

Once Dr. Nolan and his colleagues received the samples, they were able to retrieve fragments of DNA from bone marrow cells without much struggle. "We could tell this was human right away," said Atul Butte, a computational biologist at the University of California, San Francisco, and a co-author of the new study.

The scientists eventually managed to reconstruct much of Ata's genome. She was a girl, they found, most closely related to indigenous Chileans. But she also had a substantial amount of European ancestry.

The scientists have not carried out any precise dating of the skeleton, so they can't say exactly when Ata lived. But her European heritage suggested it was sometime after Chile was colonized in the 1500s.

After death, DNA disintegrates into fragments, which become smaller over the centuries. Ata's DNA fragments are still large, another clue that she's less than 500 years old.

While her elongated head was striking, it wasn't the strangest feature of Ata's skeleton. Despite being the size of a human fetus, about the length of a pen, her bones were as developed in some ways as those of a 6-year-old.

Ralph S. Lachman, an expert on hereditary bone diseases at Stanford University, examined her X-rays. He concluded that her constellation of symptoms did not match any known disease. The scientists reasoned that Ata might have had mutations for a disorder that had never before been described.

Sanchita Bhattacharya, a researcher in Dr. Butte's lab, searched for mutations in Ata's DNA and identified 2.7 million variants throughout the genome. She whittled this list to 54 rare mutations that could potentially shut down the gene in which they were located.

"I was amazed by how much you can tell from the genetic blueprint," said Ms. Bhattacharya.

Many of those genes, it turned out, are involved in building skeletons. Some have already been linked to conditions ranging from scoliosis to dwarfism to having an abnormal number of ribs.

But some of Ata's mutations are new to science. It's possible some caused her skeleton to mature quickly even while failing to grow to normal stature.

Ms. Bhattacharya speculates that such a disorder would have caused the child to be stillborn. And she stressed that these mutations are, for now, only theoretical candidates.

Other experts concurred. "There is no single slam-dunk finding that explains the bizarre appearance of this individual," said Daniel G. MacArthur, a geneticist at the Broad Institute who was not involved in the study.

Yet understanding what happened to Ata might shed light on skeletal deformities seen today. That may require engineering stem cells with each of the 54 mutations, growing them in a dish, and then looking for telling changes in their development.

And Dr. Nolan has heard stories about similar skeletons in other parts of the world. If he were able to examine them, he might discover some of these mutations in their DNA, as well.

Even more direct confirmation might be possible if researchers paid closer attention to stillbirths.

Although there are 24,000 stillbirths in the United States alone each year, doctors generally don't record the features of the fetuses, let alone study their DNA. With so little data, there's no way to know if Ata was unique.

"This could be a trigger to look into more such cases," said Albert Zink, an anthropologist at the European Research Academy in Bolzano, Italy, who was not involved in the new study.

While Dr. Nolan began the project as "a lark," he believes the evidence now requires that the mummy be returned to Chile for proper treatment as human remains.

"One has to respect these things," he said.

A Radar Blip, a Flash of Light: How U.F.O.s 'Exploded' Into Public View

BY LAURA M. HOLSON | AUG. 3, 2018

IN THE EARLY MORNING of July 20, 1952, Capt. S.C. "Casey" Pierman was ready for takeoff at Washington National Airport, when a bright light skimmed the horizon and disappeared. He did not think much of it until he was airborne, bound for Detroit, and an air traffic controller told him two or three unidentified flying objects were spotted on radar traveling at high speed.

The controller told Captain Pierman to follow them, the pilot told government investigators at the time. Captain Pierman agreed, and headed northwest over West Virginia where he saw as many as seven bluish-white lights that looked "like falling stars without tails," according to a newspaper report.

The sighting of whatever-they-were garnered headlines around the world. And in the decades since, U.F.O.s have become part of the pop culture zeitgeist, from "Close Encounters of the Third Kind" to "The X-Files." In September, a star of that long-running series, Gillian Anderson, will appear in "UFO," a movie about a college student haunted by sightings of flying saucers. A "Men in Black" remake is in the works. And the History Channel plans to air "Project Blue Book," a scripted series about the government program that studied whether U.F.O.s were a national threat.

And the topic is back in the headlines. Last year, The Times wrote about a little known project founded in 2007, the Advanced Aerospace Threat Identification Program, to investigate U.F.O. sightings. A search of The Times's historical archives reveals a rich bounty of U.F.O. sightings, lore and explanations since the 1950s. And who can forget in 2016 when Hillary Clinton said she would reopen the real X-files if she were president?

> # *Flying Objects Near Washington Spotted by Both Pilots and Radar*
>
> ## Air Force Reveals Reports of Something, Perhaps 'Saucers,' Traveling Slowly But Jumping Up and Down

THE NEW YORK TIMES

In July 1952, several U.F.O. sightings in Washington garnered headlines around the world. This one is from The New York Times.

Captain Pierman's 68-year-old daughter, Faith McClory, said in an interview last month that her father became something of a celebrity as reports like his in the summer of 1952 fueled fear of a space alien invasion.

"My sister has memories of men coming to our home," said Ms. McClory, who grew up in Belleville, Mich. (She said they were reporters.) "People were enthralled with the flying saucers," she added.

Researchers say government officials have sought to publicly debunk the existence of alien evidence ever since the Washington sightings.

"Unidentified flying objects exploded into the public consciousness then," said Mark Rodeghier, the scientific director for the Center for UFO Studies, a group of scientists and researchers who study the U.F.O. phenomenon. "There was concern in a way you hadn't seen before."

It should be noted that the term U.F.O., as used by the government, does not mean extraterrestrials from outer space. It means any object in the sky that has not been identified. When asked recently about the 1952 Washington sightings, Ann Stefanek, chief of media operations for the Air Force, wrote in an email that the objects had posed no threat to national security.

In the spring of 1952, though, numerous mysterious sightings had captured the Air Force's attention. It created "Project Blue Book" that March — the third investigative government project of its kind and the one that lasted the longest, until 1969.

The events in Washington were not the first unexplained encounter report. Debris from what observers called a "flying disc" had been spotted in Roswell, N.M., five years earlier, which Army officials said was from a "weather balloon." By 1952, though, a number of sightings of U.F.O.s were being reported across the country and the nation was on edge.

Life magazine was one of the first mainstream magazines to suggest the phenomenon was real and revealed in an April story that the Air Force was secretly investigating. That story inspired a sharp increase in reports of sightings that summer.

The Washington sightings centered on events that started around 11:40 p.m. on July 19, as air traffic controllers at Washington National Airport noticed blips speeding near Andrews Air Force Base, according to government accounts. The unidentified aircrafts fanned out, flying over the White House and the U.S. Capitol. Captain Pierman saw them that night. They vanished around 5 a.m.

It was a second sighting a week later, though, that caused the wave of hysteria that forced the government to speak out. Albert Chop, then a spokesman with the Pentagon who was given the job of answering questions about U.F.O.s, said he was awakened by a call on the evening of July 26.

Mr. Chop described the events in a 1999 oral history to the Sign Historical Group, an association of archivists and amateur researchers who held a workshop that year to study U.F.O history. He said the new objects were spotted on radar at Washington National Airport and he was told to get there right away.

The Air Force dispatched jet fighters from New Castle, Del., to intercept the flying objects. But every time one of the jets closed in, they disappeared. When the jets backed off, they reappeared.

"It was frightening," Mr. Chop said. "I think everybody in the room was very apprehensive."

At one point, a pilot found himself in the midst of four unidentified aircrafts and asked what to do. "I didn't say anything," Mr. Chop told the interviewers. "Nobody said anything. All of a sudden these things began to move away from him and he said, 'They're gone!' " The pilot returned to his base.

"These things hung around all night long," Mr. Chop added.

The next day, almost every major newspaper wrote about the U.F.O.s. " 'Objects' Outstrip Jets Over Capital," was the headline in The Times.

"People didn't know what to think of it," said Rob Swiatek, a U.F.O. researcher and scientist, in a recent interview. "They were very disturbed."

Public panic was a problem for the Air Force, which feared a diversion of resources during the Cold War. "The Air Force and the Central Intelligence Agency became worried that the Soviet Union would take advantage of the situation and launch an attack on the United States," Mr. Rodeghier said. "They were looking at worst-case scenarios."

Worse, no one could explain the phenomenon to President Harry Truman, according to press reports. One theory promoted by the Air Force was that a layer of hot air in the sky, called a temperature inversion, caused radar to mistake a weather event for flying objects. "Nobody had any answers," Mr. Chop told the interviewers. "That's why General Samford had the press conference."

On July 29, 1952, Maj. Gen. John Samford, the director of Air Force intelligence overseeing the inquiry, held a news conference at the Pentagon to reassure the public. He dismissed the Washington sightings as a temperature anomaly. Still, the general conceded that not all the details could be explained by natural causes. Witness reports "have been made by credible observers of relatively incredible things," he said at the time. "It is this group of observations that we now are attempting to resolve."

The news conference was front page news, including in The Times,

> **Air Force Debunks 'Saucers' As Just 'Natural Phenomena'**
>
> Intelligence Chief Denies a Menace Exists —'Objects' Believed to Be Reflections, but 'Adequate' Guard Will Be Kept

THE NEW YORK TIMES

After several sightings of unidentified flying objects in Washington in July 1952, government officials held a news conference, calling them "natural phenomena." The press conference was front page news, including in The Times.

which ran the headline "Air Force Debunks 'Saucers' as Just 'Natural Phenomena.'"

Case closed? Not quite.

"I don't think temperature inversion had much to do with it, but the news media accepted that explanation at the time," said Kevin Randle, a former lieutenant colonel in the Army who has studied the events of July 1952 and is the author of the 2001 book, "Invasion Washington: U.F.O.s Over the Capitol."

In January 1953, spurred by the Washington sightings, a scientific committee led by Howard Robertson, a well-known mathematician and physicist, was formed by the government to explore the phenomenon. "One of the conclusions was that they needed to debunk U.F.O.s," Mr. Randle said.

The committee, called the Robertson Panel, suggested in its report that the government conduct a mass media education campaign to "reduce the current gullibility of the public and consequently their susceptibility to clever hostile propaganda."

The campaign to re-educate Americans did not work — U.F.O.s have persisted as a fixture in pop culture. Besides, the government's explanation was not that convincing anyway. Ms. McClory, Captain Pierman's daughter, said her father did not believe that the bluish-white lights he saw were weather-related.

"I don't want to use the words 'cover up,' " she said, of her father's view. "But it was very clear. He saw it. Everything was seen on radar."

Glossary

altruistic Unselfish; showing a care or interest in the welfare of others.

anomaly Something that deviates from the norm or from what is expected.

Area 51 A top-secret air force base in rural Nevada rumored to be the location of UFO studies.

astrobiology The study of the origin, evolution, distribution, and future of life in the universe.

astrophysics The branch of astronomy that applies principles of physics and chemistry to the study of the physical nature of stars and planets.

calibration The act of checking and rectifying the gradation of an instrument designed to take quantitative measurements.

crop circle A pattern that arises from the flattening of crops; often believed to be associated with an extraterrestrial presence.

debris Scattered or loose remains of something that has been broken down or destroyed.

enthusiast A person highly interested in a subject or activity.

exoplanet A planet outside of the solar system.

fuselage The primary body of an aircraft.

hoax A plan or an act intended to deceive or to mislead a person or a group of people.

insurmountable Too great to be overcome.

metaphysical Relating to the transcendent or to what is available beyond human senses.

meteorite A meteor that survives its passage through the earth's atmosphere and makes contact with the ground.

paranormal Events and phenomena that are beyond the range of human comprehension and scientific understanding.

skeptic A person inclined to question or doubt offered and established opinions and beliefs.

telepathic Capable of transmitting thoughts to other people as well as knowing their thoughts in turn.

zeitgeist The defining spirit or mood of a particular period of history.

Media Literacy Terms

"Media literacy" refers to the ability to access, understand, critically assess and create media. The following terms are important components of media literacy, and they will help you critically engage with the articles in this title.

angle The aspect of a news story that a journalist focuses on and develops.

attribution The method by which a source is identified or by which facts and information are assigned to the person who provided them.

balance Principle of journalism that both perspectives of an argument should be presented in a fair way.

bias A disposition of prejudice in favor of a certain idea, person or perspective.

credibility The quality of being trustworthy and believable, said of a journalistic source.

critical review A type of story that describes an event or work of art, such as a theater performance, film, concert, book, restaurant, radio or television program, exhibition or musical piece, and offers critical assessment of its quality and reception.

editorial Article of opinion or interpretation.

fake news A fictional or made-up story presented in the style of a legitimate news story, intended to deceive readers; also commonly used to criticize legitimate news because of its perspective or unfavorable coverage of a subject.

feature story Article designed to entertain as well as to inform.

headline Type, usually 18 point or larger, used to introduce a story.

impartiality Principle of journalism that a story should not reflect a journalist's bias and should contain balance.

intention The motive or reason behind something, such as the publication of a news story.

news story An article or style of expository writing that reports news, generally in a straightforward fashion and without editorial comment.

op-ed An opinion piece that reflects a prominent individual's opinion on a topic of interest.

paraphrase The summary of an individual's words, with attribution, rather than a direct quotation of their exact words.

plagiarism An attempt to pass another person's work as one's own without attribution.

quotation The use of an individual's exact words indicated by the use of quotation marks and proper attribution.

reliability The quality of being dependable and accurate, said of a journalistic source.

rhetorical device Technique in writing intending to persuade the reader or communicate a message from a certain perspective.

source The origin of the information reported in journalism.

style A distinctive use of language in writing or speech; also a news or publishing organization's rules for consistent use of language with regard to spelling, punctuation, typography and capitalization, usually regimented by a house style guide.

tone A manner of expression in writing or speech.

Media Literacy Questions

1. Identify the various sources cited in the article "Glowing Auras and 'Black Money': The Pentagon's Mysterious U.F.O. Program" (on page 67). How do the journalists attribute information to each of these sources in their article? How effective are their attributions in helping the reader identify their sources?

2. In "Visitors From Outer Space, Real or Not, Are Focus of Discussion in Washington" (on page 52), Andrew Siddons directly quotes Mike Gravel. What are the strengths of the use of a direct quote as opposed to a paraphrase? What are the weaknesses?

3. Compare the headlines of "Bright Lights, Strange Shapes and Talk of U.F.O.'s" (on page 194) and "Twinkle, Twinkle Little Trappist" (on page 96). Which is a more compelling headline, and why? How could the less compelling headline be changed to better draw the reader's interest?

4. The article "Flying Saucers and Other Fairy Tales" (on page 34) is an example of an op-ed. Identify how Ross Douthat's attitude and tone help convey his opinion on the topic.

5. Does "C.I.A. Acknowledges Area 51 Exists, but What About Those Little Green Men?" (on page 56) use multiple sources? What are the strengths of using multiple sources in a journalistic piece? What are the weaknesses of relying heavily on only one or a few sources?

6. What is the intention of the article "A U.F.O., or Is There an Explanation?" (on page 186)? How effectively does it achieve its intended purpose?

7. Analyze the authors' reporting in "Taking U.F.O.'s for Credit, and for Real" (on page 14) and "U.F.O. Believers and Debunkers Thrive on the Web" (on page 17). Do you think one journalist is more balanced in his reporting than the other? If so, why do you think so?

8. Identify each of the sources in "Pit Stop for U.F.O.'s, and Humans Who Love Them" (on page 26) as a primary source or a secondary source. Evaluate the reliability and credibility of each source. How does your evaluation of each source change your perspective on this article?

9. Do Helene Cooper, Leslie Kean and Ralph Blumenthal demonstrate the journalistic principle of impartiality in their article "2 Navy Airmen and an Object That 'Accelerated Like Nothing I've Ever Seen' " (on page 64)? If so, how did they do so? If not, what could they have included to make their article more impartial?

10. "They've 'Seen Things' " (on page 42) features photographs of a group of U.F.O. enthusiasts. What do these photographs add to the article?

Citations

All citations in this list are formatted according to the Modern Language Association's (MLA) style guide.

BOOK CITATION

THE NEW YORK TIMES EDITORIAL STAFF. *Extraterrestrials and U.F.O.s*. New York: New York Times Educational Publishing, 2020.

ONLINE ARTICLE CITATIONS

BLUMENTHAL, RALPH. "On the Trail of a Secret Pentagon U.F.O. Program." *The New York Times*, 18 Dec. 2017, https://www.nytimes.com/2017/12/18/insider/secret-pentagon-ufo-program.html.

BLUMENTHAL, RALPH. "People Are Seeing U.F.O.s Everywhere, and This Book Proves It."*The New York Times*, 24 Apr. 2017, https://www.nytimes.com/2017/04/24/science/ufo-sightings-book.html.

BROMWICH, JONAH ENGEL. "Bright Lights, Strange Shapes and Talk of U.F.O.'s." *The New York Times*, 12 Nov. 2015, https://www.nytimes.com/2015/11/13/us/bright-lights-strange-shapes-and-talk-of-ufos.html.

CHANG, KENNETH. "A Nearby Earth-Size Planet May Have Conditions for Life." *The New York Times*, 15 Nov. 2017, https://www.nytimes.com/2017/11/15/science/planet-ross-128.html.

CHANG, KENNETH, AND DENNIS OVERBYE. "A Large Body of Water on Mars Is Detected, Raising the Potential for Alien Life." *The New York Times*, 25 July 2018, https://www.nytimes.com/2018/07/25/science/mars-liquid-alien-life.html.

CHOZICK, AMY. "Hillary Clinton Gives U.F.O. Buffs Hope She Will Open the X-Files." *The New York Times*, 10 May 2016, https://www.nytimes.com/2016/05/11/us/politics/hillary-clinton-aliens.html.

COOPER, HELENE, ET AL. "Glowing Auras and 'Black Money': The Pentagon's Mysterious U.F.O. Program." *The New York Times*, 16 Dec. 2017, https://www.nytimes.com/2017/12/16/us/politics/pentagon-program-ufo-harry-reid.html.

COOPER, HELENE, ET AL. "2 Navy Airmen and an Object That 'Accelerated Like Nothing I've Ever Seen.' " *The New York Times*, 16 Dec. 2017, https://www.nytimes.com/2017/12/16/us/politics/unidentified-flying-object-navy.html.

DOUTHAT, ROSS. "Flying Saucers and Other Fairy Tales." *The New York Times*, 23 Dec. 2017, https://www.nytimes.com/2017/12/23/opinion/alien-encounters-christmas-ufo.html.

FOX, MARGALIT. "Betty Hill, 85, Figure in Alien Abduction Case, Dies." *The New York Times*, 23 Oct. 2004, https://www.nytimes.com/2004/10/23/us/betty-hill-85-figure-in-alien-abduction-case-dies.html.

FOX, MARGALIT. "Budd Hopkins, Abstract Expressionist and U.F.O. Author, Dies at 80." *The New York Times*, 24 Aug. 2011, https://www.nytimes.com/2011/08/25/arts/design/budd-hopkins-abstract-artist-and-ufo-author-dies-at-80.html.

FRANK, ADAM. "Yes, There Have Been Aliens." *The New York Times*, 10 June 2016, https://www.nytimes.com/2016/06/12/opinion/sunday/yes-there-have-been-aliens.html.

DE FREYTAS-TAMURA, KIMIKO. "Winston Churchill Wrote of Alien Life in a Lost Essay." *The New York Times*, 15 Feb. 2017, https://www.nytimes.com/2017/02/15/world/europe/winston-churchill-aliens.html.

GOODMAN, J. DAVID. "Formations in China Desert Are Still a Mystery." *The New York Times*, 18 Nov. 2011, https://thelede.blogs.nytimes.com/2011/11/18/formations-in-china-desert-are-still-a-mystery/.

GRIMES, WILLIAM. "Ionel Talpazan, Whose U.F.O. Art Had Sightings All Over, Dies at 60." *The New York Times*, 29 Sept. 2015, https://www.nytimes.com/2015/09/30/arts/design/ionel-talpazan-whose-ufo-art-had-sightings-all-over-dies-at-60.html.

HINDS, MICHAEL DECOURCY. "Campus Journal; Taking U.F.O.'s for Credit, and for Real." *The New York Times*, 28 Oct. 1992, https://www.nytimes.com/1992/10/28/news/campus-journal-taking-ufo-s-for-credit-and-for-real.html.

HOLSON, LAURA J. "A Radar Blip, a Flash of Light: How U.F.O.s 'Exploded' Into Public View." *The New York Times*, 3 Aug. 2018, https://www.nytimes.com/2018/08/03/science/UFO-sightings-USA.html.

JOHNSON, GEORGE. "The Intelligent-Life Lottery." *The New York Times*, 18 Aug. 2014, https://www.nytimes.com/2014/08/19/science/in-search-for-intelligent-life-consider-the-lottery.html.

JOHNSON, GEORGE. "Why We Keep Dreaming of Little Green Men." *The New York Times*, 13 May 2016, https://www.nytimes.com/2016/05/15/opinion/sunday/why-we-keep-dreaming-of-little-green-men.html.

JOHNSON, KIRK. "Pit Stop for U.F.O.'s, and Humans Who Love Them." *The New York Times*, 25 Nov. 2010, https://www.nytimes.com/2010/11/26/us/26ufo.html.

JOHNSON, STEVEN. "Greetings, E.T. (Please Don't Murder Us.)." *The New York Times*, 28 June 2017, https://www.nytimes.com/2017/06/28/magazine/greetings-et-please-dont-murder-us.html.

JONES, CHRIS. "The Woman Who Might Find Us Another Earth." *The New York Times*, 7 Dec. 2016, https://www.nytimes.com/2016/12/07/magazine/the-world-sees-me-as-the-one-who-will-find-another-earth.html.

LYONS, PATRICK J. "U.F.O. Believers and Debunkers Thrive on the Web." *The New York Times*, 30 June 1997, https://www.nytimes.com/1997/06/30/business/ufo-believers-and-debunkers-thrive-on-the-web.html.

MARTIN, DOUGLAS. "Philip Klass, 85, Debunker of Claims of Flying Saucers, Dies." *The New York Times*, 12 Aug. 2005, https://www.nytimes.com/2005/08/12/us/philip-klass-85-debunker-of-claims-of-flying-saucers-dies.html.

NAGOURNEY, ADAM. "C.I.A. Acknowledges Area 51 Exists, but What About Those Little Green Men?" *The New York Times*, 22 Aug. 2013, https://www.nytimes.com/2013/08/23/us/cia-acknowledges-area-51-exists-but-what-about-those-little-green-men.html.

THE NEW YORK TIMES. "Twinkle, Twinkle Little Trappist." *The New York Times*, 24 Feb. 2017, https://www.nytimes.com/2017/02/24/opinion/twinkle-twinkle-little-trappist.html.

NIZZA, MIKE. "When an Astronaut Believes in Aliens." *The New York Times*, 24 July 2008, https://thelede.blogs.nytimes.com/2008/07/24/when-an-astronaut-believes-in-aliens/.

OVERBYE, DENNIS. "The Flip Side of Optimism About Life on Other Planets." *The New York Times*, 3 Aug. 2015, https://www.nytimes.com/2015/08/04/science/space/the-flip-side-of-optimism-about-life-on-other-planets.html.

OVERBYE, DENNIS. "An Interstellar Visitor Both Familiar and Alien." *The New York Times*, 22 Nov. 2017, https://www.nytimes.com/2017/11/22/science/oumuamua-space-asteroid.html.

OVERBYE, DENNIS. "A New Exoplanet May Be Most Promising Yet in Search for Life." *The New York Times*, 19 Apr. 2017, https://www.nytimes.com/2017/04/19/science/exoplanet-signs-of-life.html.

OVERBYE, DENNIS. "U.F.O.s: Is This All There Is?" *The New York Times*, 29 Dec. 2017, https://www.nytimes.com/2017/12/29/science/ufos-aliens-space-travel.html.

PARISI, ALBERT J. "A U.F.O., or Is There an Explanation?" *The New York Times*, 1 Jan. 1995, https://www.nytimes.com/1995/01/01/nyregion/a-ufo-or-is-there-an-explanation.html.

RAGO, ROZETTE. "They've 'Seen Things.'" *The New York Times*, 14 Aug. 2018, https://www.nytimes.com/2018/08/14/us/theyve-seen-things.html.

REGENOLD, STEPHEN. "Lonesome Highway to Another World?" *The New York Times*, 13 Apr. 2007, https://www.nytimes.com/2007/04/13/travel/escapes/13extraterrestrial.html.

RYERSON, JAMES. "The Truth Is Out There." *The New York Times*, 23 Sept. 2016, https://www.nytimes.com/2016/09/25/books/review/search-for-extraterrestrial-intelligence.html.

SANDOMIR, RICHARD. "Ursula Marvin, Geologist of the Extraterrestrial, Dies at 96." *The New York Times*, 9 Mar. 2018, https://www.nytimes.com/2018/03/09/obituaries/ursula-marvin-geologist-of-the-extraterrestrial-dies-at-96.html.

SENIOR, JENNIFER. "'Aliens' Asks: If the Universe Is So Vast, Where Is Everybody?" *The New York Times*, 24 May 2017, https://www.nytimes.com/2017/05/24/books/review-aliens-search-for-extraterrestrial-life-edited-jim-al-khalili.html.

SHOSTAK, SETH. "Should We Keep a Low Profile in Space?" *The New York Times*, 27 Mar. 2015, https://www.nytimes.com/2015/03/28/opinion/sunday/messaging-the-stars.html.

SIDDONS, ANDREW. "Visitors From Outer Space, Real or Not, Are Focus of Discussion in Washington." *The New York Times*, 3 May 2013, https://www.nytimes.com/2013/05/04/us/politics/panel-convenes-in-washington-to-discuss-aliens.html.

SULLIVAN, WALTER. "Despite Lack of Data From Pilots and Officials, Reports of UFO Sightings Are Many and Widespread." *The New York Times*, 21 Oct. 1973, https://www.nytimes.com/1973/10/21/archives/despite-lack-of-data-from-pilots-and-officials-reports-of-ufo.html.

YARDLEY, WILLIAM. "John Billingham, Seeker of Extraterrestrials, Dies at 83." *The New York Times*, 10 Aug. 2013, https://www.nytimes.com/2013/08/11/science/space/john-billingham-seeker-of-extraterrestrials-dies-at-83.html.

ZELLER, TOM, JR. "U.F.O. Hits Congestion at O'Hare, Turns Back." *The New York Times*, 2 Jan. 2007, https://thelede.blogs.nytimes.com/2007/01/02/ufo-hits-congestion-at-ohare-turns-back.

ZIMMER, CARL. "Was a Tiny Mummy in the Atacama an Alien? No, but the Real Story Is Almost as Strange." *The New York Times*, 22 Mar. 2018, https://www.nytimes.com/2018/03/22/science/ata-mummy-alien-chile.html.

Index

A

Advanced Aerospace Threat Identification Program, 37, 44, 67–74, 75–77, 205
"Aliens: The World's Leading Scientists on the Search for Extraterrestrial Life," 101–104
Allen Telescopic Array, 80, 152
All These Worlds Are Yours: The Scientific Search for Alien Life, 32
Alpha Centauri, 40, 88, 159
Area 51, 10–13, 17, 20, 23, 24, 52, 56–59, 60, 192
Arecibo message, 105–106, 111–114, 116, 123, 127
Arnold, Kenneth, 179
Arnu, Joerg, 20, 24–25
artworks, inspired by extraterrestrials, 147, 154–156
Atacama, 201–204
Atomic Energy Commission's Nevada Test Site, 11
Australian Skeptics, 18

B

Bigelow, Robert, 67, 68, 71, 72, 76
Bigelow Aerospace, 72
Billingham, John, 151–153
Bingham, Robert, 42–43, 45–49
Bioastronomy News, 78
Blumenthal, Ralph, 64–66, 67–74, 75–77, 197–200
Bostrom, Nick, 87, 89, 90
"Boyajian's star," 40
Breakthrough Listen, 107, 108
Bromwich, Jonah Engel, 194–196

C

Cape Town, 194
Central Intelligence Agency (C.I.A.), 56–59, 69, 70, 72, 76, 191
"The Central Intelligence Agency and Overhead Reconnaissance: The U-2 and Oxcart Programs, 1954-1974," 48
Chang, Kenneth, 132–134, 135–139
"Chariots of the Gods," 34–35
Chile, 33, 98, 100, 132, 201–204
Chozick, Amy, 60–63
Churchill, Winston, 171–174
Circlemakers, 19
Citizen Hearing on Disclosure, 52–55
Clinton, Hillary, 10, 13, 60–63, 205
Coar, Robert F., 186–188
college courses, on extraterrestrials, 14–16
Committee for Scientific Investigation of Claims of the Paranormal (Csicop), 18
Cooper, Helene, 64–66, 67–74
crop circles, 19, 192

D

"Darwin's Dangerous Idea," 81
debunkers/debunking, 17–19, 50, 69, 77, 143–144, 206, 209
"Declaration of Principles Concerning Activities Following the Detection of Extraterrestrial Intelligence," 153
De Freytas-Tamura, Kimiko, 171–174
DeLonge, Tom, 37, 45
Dennett, Daniel, 81
Denning, Kathryn, 124–126
Drake, Frank, 8, 31, 82, 90, 92, 105, 106, 107, 110, 111–114, 116, 119, 120–122, 123, 127, 151
Drake equation, 92, 94–95, 120–122
Douthat, Ross, 34–36

E

Easton, James, 19
Einstein, Albert, 38, 94

Elizondo, Luis, 43, 45, 67, 69, 72, 73, 74, 75, 76
Enceladus, 32, 136
Europa, 32, 87, 136
European Extremely Large telescope, 100
exoplanets, 30, 32, 33, 92, 96–97, 98–100, 102, 111, 114, 158–161, 163, 166, 167

F
fast radio bursts, 40
Fermi, Enrico, 88, 90, 101
Fermi Paradox, 89, 101, 120, 123
51 Pegasi b, 159
Fox, Margalit, 140–142, 147–150
Frank, Adam, 92, 94–95
Fravor, David, 37, 41, 64–66
Freudenthal, Hans, 115
Fuller, John G., 140
Fuller, Paul, 19

G
Gertz, John, 119
Giant Magellan telescope, 100
Gliese 411, 114
Gobi Desert, 191–193
Goldilocks zone, 98, 110, 114, 115, 120, 158
Goodman, J. David, 191–193
Gould, Stephen Jay, 81
Great Filter, 89
Grimes, William, 154–156
Guccione, Bob, 19

H
Hangar 18, 17
Harps spectography, 98
Hawking, Stephen, 84, 102, 108, 117, 119, 122, 123

Hill, Betty, 140–142
Hinds, Michael Decourcy, 14–16
Holson, Laura M., 205–210
Hopkins, Budd, 147–150

I
Indigo Army, 47, 48
International U.F.O. Museum and Research Center, 17
"The Interrupted Journey: Two Lost Hours 'Aboard a Flying Saucer,'" 140
"Intruders: The Incredible Visitations at Copley Woods," 149

J
Jacobs, David, 14–16
James Webb Space Telescope, 100
Johnson, George, 10–13, 78, 80–82
Johnson, Kirk, 26–29
Johnson, Steven, 105–131
Jones, Chris, 157–170

K
Kean, Leslie, 64–66, 67–74
Kepler 452b, 87
Kepler-186f, 166, 168
Kepler-16b, 166
Klass, Philip, 19, 143–144

L
L.A. U.F.O. Channel, 46
LHS 1140, 98–100
"Lincos: Design of a Language for Cosmic Intercourse," 115
Little A'Le'Inn, 23, 24, 56–59
Lyman, Spencer, 161, 162
Lyons, Patrick J., 17–19

M
Manikin Who Fell to Earth, 19
Mars, discovery of large body of water on, 8, 78, 135–139
Martin, Douglas, 143–144
Marvin, Ursula, 175–178
Mayr, Ernst, 78, 79, 80, 81
Messeri, Lisa, 32–33
Messoline, Judy, 26–29
METI (Messaging Extra Terrestrial Intelligence), 107–108, 110, 114–126
Milner, Yuri, 87, 107, 108
Mitchell, Edward, 145–146
Moseley, James, 19
Mutual U.F.O. Network, (Mufon) 17, 43

N
Nagourney, Adam, 56–59
National U.F.O. Reporting Center, 17
Nevada State Route 375 (Extraterrestrial Highway), 20–25
Nizza, Mike, 145–146

O
Obama, Barack, 12, 52, 54, 60, 194
Office of the Search for Extraterrestrial Intelligence, 153
O'Hare airport, extraterrestrial sighting at, 189–190
Olbers' Paradox, 90
100-Year Starship Study, 89
Oumuamua, 40, 78, 128–131
Overbye, Dennis, 37–41,

INDEX **223**

87–91, 98–100, 128–131, 135–139

P

Pan-STARRS 1 telescope, 128, 129, 130
Paradigm Research Group, 53
Parascope, 19
Parisi, Albert, 186–188
"Passport to Magonia: From Folklore to Flying Saucers," 34–35
Peckman, Jeff, 145
Pentagon, 37, 43, 54, 64, 67–74, 75–77
Pierman, S.C., 205–210
Placing Outer Space: An Earthly Ethnography of Other Worlds, 32–33
Podesta, John, 54, 55, 61, 62
Project Blue Book, 41, 70, 207
Project Cyclops, 152
Project Galactic Guide, 19
Proxima b, 30, 99, 160, 166
PSO J318.5-22, 166

R

Rago, Rozette, 42–51
Randi, James, 18
Regenold, Stephen, 20–25
Reid, Harry, 34, 67, 68, 71, 73, 74, 75, 76
Rockefeller, Laurance S., 62, 63
Ross 128, 132–134
Roswell, 17, 19, 34, 63, 50, 207
Roswell Museum, 19
"Roswell Report," 18
"The Roswell Report: Case Closed," 17–18

Ryerson, James, 30–33
Ryle, Martin, 106, 108

S

Sagan, Carl, 8, 78, 79, 80, 90, 107, 110, 120, 152
Sandomir, Richard, 175–178
Saucer Smear, 19
Seager, Sara, 44, 69, 99, 157–170
"Secret Life: Firsthand Accounts of UFO Abductions," 14–15, 16
Senior, Jennifer, 101–104
SETI (Search for Extraterrestrial Intelligence), 8, 30–33, 40, 78, 80–82, 83–86, 103, 107, 110, 115, 119, 120, 122, 124, 151, 152, 153
Shostak, Seth, 80, 83–86, 103, 115–116
Siddons, Andrew, 52–55
Skeptical Inquirer, 18
Slaight, Jim, 64–66
Sleeping ET Quiz, 19
Sputnik IV, 175
Squeri, Lawrence, 30–31
starshade, 161–162
Strieber, Whitley, 142, 148
Sullivan, Walter, 179–185
Sullivan, Woodruff, 92, 95

T

Talpazan, Ionel, 154–156
Tarter, Jill, 40
"Temple of Apollo With Guardian XXXXV," 147
Temple University, 14–16
Titan, 32
To the Stars Academy of Arts & Science, 37–38, 45, 74, 76
Trappist-1, 96–97, 99, 111

Truly Dangerous, 19
Trump, Donald, 10
Tyson, Neil deGrasse, 31–32

U

"UFO Abductions: A Dangerous Game," 143
U.F.O. and Paranormal Research Society, 43
Uforia, 17
"U.F.O. Sightings Desk Reference,"197–200
U.F.O. Watchtower, 26–29
U-2 spy plane, 11, 56, 58

V

Vakoch, Doug, 107–110, 114, 115, 116, 117, 118, 119, 123, 125
Vallée, Jacques, 34–35
"Visions of Space & UFOs in Art," 155
von Däniken, Erich, 34–35

W

"Waiting for Contact: The Search for Extraterrestrial Intelligence," 30–31
"Welcome to the Universe: An Astrophysical Tour," 31–32
Willis, John, 32
"Wonderful Life," 81

Y

Yardley, William, 151–153

Z

Zeller, Tom, Jr., 189–190
Zimmer, Carl, 201–204
Zond IV, 185
zoo hypothesis, 40

This book is current up until the time of printing. For the most up-to-date reporting, visit www.nytimes.com.